SUGAR-BEET NUTRITION

Sugar-Beet Nutrition

A. P. DRAYCOTT
B.Sc., Ph.D.

Head of the Chemistry Section
Broom's Barn Experimental Station
Bury St. Edmunds, Suffolk, England

APPLIED SCIENCE PUBLISHERS LTD
LONDON

APPLIED SCIENCE PUBLISHERS LTD
RIPPLE ROAD, BARKING, ESSEX, ENGLAND

ISBN: 0 85334 550 3

WITH 29 ILLUSTRATIONS AND 100 TABLES

Printed in Great Britain by Galliard Limited, Great Yarmouth, Norfolk, England

Preface

The justification for attempting a monograph on the nutrient requirements of sugar beet is the complete absence of a comprehensive, up-to-date account of the subject in the English language. Sugar beet supplies nearly half the world's requirement of sugar and the crop occupies over 19 million acres each year. In most countries, fertiliser is the most expensive item in the variable costs of growing the crop. For example, in Great Britain the approximate cost of production per acre without hand labour is: seed, £3.40; herbicide and insecticide sprays, £8.00; haulage, £7.50; and fertiliser, £11.30.

In Great Britain the sugar-beet crop occupies 450 000 acres or 4% of the arable area and growers spend over £5 million on fertiliser for the crop annually. To ensure that this fertiliser is used wisely, the Sugar Beet Research and Education Committee of the Ministry of Agriculture has financed experiments with fertilisers for the sugar-beet crop for nearly 40 years. Initially the experiments were co-ordinated by staff at Rothamsted Experimental Station (1933–49), notably the late Dr E. M. Crowther. From 1956–1961 experiments were organised from Dunholme Field Station by Dr S. N. Adams and latterly from Broom's Barn Experimental Station by Dr P. B. H. Tinker (1962–65) and the writer (1965 to date).

At present the results of these investigations are scattered through many published and unpublished reports and in papers in numerous scientific journals and they are therefore not readily accessible to many of the people who could make most use of them; this book brings this experimental evidence together in one place for the first time. This does not mean to say that other experimental work on sugar-beet nutrition in other countries has been ignored for, wherever possible, the results are set in the context of published data from other sources. Much material about the residual effects of fertilisers on sugar beet has also been obtained from the results of classical and long-term experiments at Rothamsted, Woburn, Saxmundham and, more recently, at Broom's Barn. Where information on a topic was lacking from the British experiments, foreign evidence, particularly from the USA, has been used.

I hope that this book will serve a need, both as a reference to the present state of our knowledge on the elements needed by the crop and as a guide for farmers, advisers and research workers who are concerned with growing sugar beet.

ACKNOWLEDGEMENTS

I owe a considerable debt of gratitude to Dr R. Hull for his encouragement and help in preparing this book. My appreciation of plant nutrition and soil fertility is due in no short measure to Dr G. W. Cooke and it has been a great help to draw on his experience in writing this text. I am indebted to Mr O. S. Rose of British Sugar Corporation for permission to include the fieldmen's records of fertiliser usage and acreages affected by nutrient deficiencies; these statistics are unique to the sugar-beet crop and it is largely due to his far-sighted planning that such records are available.

I also thank my colleagues, M. J. Durrant, R. F. Farley, P. J. Last and P. C. Longden for reading the manuscript, Mrs. Joan Chapman for help with the references and index, and drawing the diagrams, and Mrs. Greta Fuller-Rowell for typing the manuscript.

A. P. DRAYCOTT

Broom's Barn Experimental Station
Higham
Bury St. Edmunds
Suffolk

April, 1972

Contents

response—soil acidity—determination of lime requirement—
factory waste lime

Uptake—deficiency symptoms—quantity in the crop—soil
magnesium—other sources of magnesium—soil analysis—
effects of fertilisers on yield—comparisons of forms and
methods of application—interactions—time of application

BORON: deficiency symptoms—anatomical effects of de-
ficiency—influence of soil and weather—concentration in the
crop—response—uptake—residual effects—plant analysis—
soil analysis
CHLORINE
COPPER: deficiency symptoms—concentration in the plant—in
soil
IRON: deficiency symptoms—response
MANGANESE: deficiency symptoms—movement in the plant—
response—effects of other fertilisers on manganese deficiency
MOLYBDENUM
RUBIDIUM
ZINC

ORGANIC MANURES: fertiliser equivalents of farmyard manure—
optimal dressings with farmyard manure—effects of other
organic manures—value other than as a supply of major
nutrients
GREEN MANURING: effects on sugar beet—value in farming
practice

Sugar beet in the rotation—residual effects and crop residues—
manurial value of sugar-beet tops—long-term experiments

Previous cropping—sodium—compaction—deep ploughing
and subsoiling—growing without ploughing—unstable soil
structure—soil moisture and emergence—optimum porosity—
harvesting damage—soil 'conditioners'

Chapter 1

Introduction

Sugar-beet cultivation

Sugar beet is a specialised type of *Beta vulgaris* grown for sugar production. It was developed in Europe at the end of the eighteenth century from white fodder beet, which was found to be the most suitable alternative source of sugar to tropical sugar cane. It is a biennial plant which stores up reserves in the root during the first growing season so that it is able to over-winter and produce flowering stems and seed in the following summer.

The sugar-beet crop is cultivated successfully in a wide range of climates on many different soils. Most is grown at latitudes between 30 and 60°N, as a summer crop in maritime, prairie and semi-continental climates and as winter or summer crop in Mediterranean and semi-arid conditions. The crop is grown with supplementary irrigation in regions where low rainfall previously prevented its cultivation.

Not only is sugar beet grown under a wide range of climates but the soils where the crop is cultivated also vary greatly. However, they are all *arable* soils, some of which have been cultivated for only a few years but many have been in arable cultivation for centuries. Soils which are cultivated and cropped continuously have many features in common, particularly in relation to their supply of the major nutrients required by sugar beet and other crops.

Nitrogen is in short supply in nearly all arable soils and it is the most important element for sugar beet in fertiliser wherever the crop is grown. When soils are first brought into intensive farming, phosphorus is usually the first fertiliser needed but many old arable soils now contain large reserves of phosphorus, residues from continual use of fertiliser; fresh phosphorus fertiliser increases sugar-beet yield little on these soils. However, despite much use of potassium fertiliser, sugar-beet yield is usually increased greatly by further applications of the element and sometimes by other cations supplied in fertiliser.

1

Amount of fertiliser used

Table 1 gives the area and average yield of sugar beet in each country and where available, the average dressing of the three major elements. Yields vary greatly from country to country but the amounts of fertiliser applied are remarkably similar. Reports from many countries suggest that the amounts of fertiliser used for sugar beet are increasing and the changes which have taken place in Great Britain over the past 30 years typify these trends.

Surveys of fertiliser use for sugar beet

ADAS/FMA/ROTHAMSTED SURVEYS

Since 1941, information about fertilisers used on crops in England and Wales has been collected in a series of surveys on representative farms by the Agricultural Development and Advisory Service (ADAS), formerly the National Agricultural Advisory Service (NAAS), latterly assisted by the Fertiliser Manufacturers' Association (FMA). Staff of the Statistics Department at Rothamsted have co-ordinated and reported on the surveys.

One of the earliest reports[382] showed that the average fertiliser usage throughout Britain in 1942/43 was 4·3 ton/acre farmyard manure, 0·5 cwt/acre N, 0·6 cwt/acre P_2O_5 and 0·3 cwt/acre K_2O. Church[56] reported the average dressings in 1945 were 0·8 cwt/acre N, 0·8 cwt/acre P_2O_5 and 0·7 cwt/acre K_2O; and 0·95 cwt/acre N, 0·95 cwt/acre P_2O_5 and 1·22 cwt/acre K_2O in 1950. Thus dressings had increased greatly compared with the early 1940's and also increased during the five-year period 1945–50. Boyd[34] showed that by 1957 the average dressings had increased to 1·1 cwt/acre N, 0·95 cwt/acre P_2O_5 and 1·6 cwt/acre K_2O. On most farms, similar quantities of the same compound fertiliser were applied for sugar beet and for potatoes. Comparing the amounts applied for sugar beet with the recommended optima, the average dressing of phosphorus of 0·95 cwt/acre P_2O_5 exceeded requirement by about 0·45 cwt/acre. Nitrogen and potassium usage were very near the optima and left little scope for improvement. The explanation put forward for the excessive usage of phosphorus was the unwillingness of most farmers to use several compounds on the farm, for most tended to use the same compound for sugar beet as that used for potatoes. Boyd et al.[35] found that 16% of the farmyard manure produced on arable farms was applied to sugar beet, which allowed 38% of the total acreage to be treated.

Church and Webber[57] recently reported on a new type of fertiliser

TABLE 1

AREA AND YIELD OF SUGAR BEET AND ESTIMATES OF THE
AMOUNT OF FERTILISER USED FOR THE CROP IN EACH
COUNTRY

	$Area^a$ 1 000 (acres)	$Yield^a$ (ton/acre)	N	P_2O_5 (cwt/acre)	K_2O	References
Afghanistan	12	5·5	—	—	—	
Albania	17	8·3	—	—	—	
Algeria	10	8·0	1·04	0·82	0·87	Bassereau[17]
Austria	116	17·0	1·04	1·01	1·21	Graf[135]
Belgium	220	19·5	1·32	1·01	2·22	Roussel[293]
Bulgaria	146	11·2	—	—	—	
Canada	79	12·1	1·20	0·82	1·59	
Chile	67	15·6	1·28	1·01	0·82	
China	563	8·8	—	—	—	
Czechoslovakia	447	12·8	1·12	0·73	1·69	Fieldler[119]
Denmark	128	15·4	1·12	0·37	1·59	Oien[260]
Finland	35	10·1	1·00	2·29	2·02	Brummer[45]
France	990	17·8	1·20	0·82	1·59	Boiteau[29]
Germany (East)	474	10·1	1·44	0·82	1·93	
Germany (West)	728	18·1	1·75	1·01	2·22	Rid[287]
Greece	54	18·8	1·20	0·82	0·38	
Hungary	239	13·5	1·12	0·64	0·72	
Iran	383	8·8	—	—	—	
Iraq	5	7·6	—	—	—	
Ireland	62	14·7	0·80	2·29	3·04	Gallagher[124]
Israel	12	18·5	1·44	1·10	1·21	Cohen[58]
Italy	719	14·5	0·80	1·01	0·82	Zocco[385]
Japan	146	14·1	1·12	0·55	1·45	
Lebanon	5	17·0	—	—	—	
Morocco	79	11·2	—	—	—	
Netherlands	254	19·3	1·12	0·73	1·11	Jorritsma[200]
Pakistan	25	6·8	0·64	—	—	
Poland	1 012	11·0	1·12	1·46	1·59	
Portugal (Azores)	9	11·5	—	—	—	
Roumania	464	8·0	0·80	0·82	0·96	Petrescu et al.[276]
Spain	449	10·9	1·00	0·73	0·82	Ontañon[263]
Sweden	99	14·7	1·16	0·55	1·06	Grönevik[138]
Switzerland	22	17·4	0·80	0·73	1·59	Meyer[244]
Syria	17	10·6	—	—	—	
Tunisia	7	6·1	0·80	0·73	0·58	Capitaine[49]
Turkey	254	13·0	0·88	0·82	0·19	Güray[139]
UK	454	13·6	1·32	0·92	1·54	
Uruguay	44	9·0	—	—	—	
USA	1 538	16·1	1·20	0·64	0·48	Hills and Ulrich[180]
USSR	8 355	8·4	0·64	—	—	
Yugoslavia	237	15·1	1·20	1·19	0·96	Markovic and Stojanovic[238]

^{a} F.A.O. Statistics for 1969.

survey begun in 1969. Farms were taken systematically to represent the whole of England and Wales not, as previously, to represent small well-defined areas. Table 2 shows the fertiliser practice in 1969. Comparisons are also made in Table 2 between average practice and recommendations appropriate to conditions under which the crop is commonly grown. Despite much advisory supervision, it receives about 30% more nitrogen and potassium and nearly double the phosphorus recommended, even without allowing for the nutrients

TABLE 2

FERTILISERS APPLIED FOR SUGAR BEET IN GREAT BRITAIN, 1969, COMPARED WITH RECOMMENDED DRESSINGS
(after Church and Webber[57])

	Nitrogen	Phosphorus	Potassium	Sodium	Lime	FYM
Area receiving treatment (%)	100	100	100	37	19	28
			Dressings			
	N	P_2O_5	K_2O	NaCl	CaO	FYM
		(cwt/acre)			(ton/acre)	
Applied	1·30	0·93	1·57	3·60	1·33	15
Recommended	1·00	0·50	1·00	3·00		
	N	P	K	Na	Ca	FYM
		(kg/ha)			(t/ha)	
Applied	163	51	166	180	3·34	37·7
Recommended	126	27	104	150		

applied in farmyard manure. Almost 40% of the crop was given between 1·20 to 1·40 cwt/acre N; use of phosphorus and potassium was more variable, usually ranging from 0·60 to 1·40 cwt/acre P_2O_5 and 1·40 to 2·00 cwt/acre K_2O, but nearly always more than the general recommendations. For a third of the crop on which sodium was also used, only 0·50 cwt/acre K_2O is recommended.

BRITISH SUGAR CORPORATION SURVEY

Fieldmen of the British Sugar Corporation have reported each year since 1957 on the amount of each of the major nutrients used on the crop and the acreage treated. Sugar beet is thus unique amongst crops, for the results provide detailed fertiliser statistics for the whole crop acreage. Table 3 summarises the usage of nitrogen, phosphorus and potassium for four-year periods from 1957 onwards (from 1965 farmers with three acres or less have been omitted). The acreage receiving each element in this period was very nearly 100%. Growers

TABLE 3

FERTILISER USAGE ON SUGAR BEET, 1957–70
(from British Sugar Corporation's fieldmen's reports)

	1957–60	1961–64	1965–68	1969–70
		(cwt/acre)		
N	0·98	1·10	1·19	1·26
P_2O_5	0·95	0·95	0·92	0·92
K_2O^a	1·52	1·53	1·33	1·32
		(kg/ha)		
N	123	138	150	158
P	52	52	51	51
K^a	158	159	138	138

^{a} Excluding kainit.

have consistently increased nitrogen dressings by 0·02–0·03 cwt/acre/annum from 0·95 cwt/acre to 1·25 cwt/acre N and it is difficult to understand why they have done so. It may in part be because of increased concentration of nitrogen in fertilisers but, more likely, farmers like to see the crop looking well and nitrogen makes the tops grow large and green. Whatever the reason, there is little experimental evidence to support the increase (see Chapter 2).

Unlike nitrogen, phosphorus usage has been remarkably stable at 0·90 to 0·95 cwt/acre P_2O_5 during the same period. Experimental evidence indicates that about half this amount would be sufficient for maximum yield (Chapter 3). The amount of potassium applied has fluctuated slightly during this period but appears to be declining slightly; if used with sodium, the present dressing would be adequate (Chapter 4). Table 4 shows, however, that only one-third to one-half of the sugar-beet acreage receives sodium each year as kainit or agricultural salt. The dressing of kainit (5·5 cwt/acre) given to

TABLE 4

AMOUNT OF AGRICULTURAL SALT AND KAINIT USED AND
AREA RECEIVING THEM, 1957–70
(from British Sugar Corporation's fieldmen's reports)

	Agricultural salt				Kainit			
	Dressing		Area ($\times 1\,000$)		Dressing		Area ($\times 1\,000$)	
	NaCl (cwt/acre)	Na (kg/ha)	(acres)	(ha)	(cwt/acre)	(kg/ha)	(acres)	(ha)
1957–60	5·1	252	37	15	5·8	728	69	28
1961–64	4·2	210	67	27	5·5	690	93	38
1965–68	4·0	200	69	28	5·2	653	107	43
1969–70	4·2	210	67	27	5·5	690	110	45

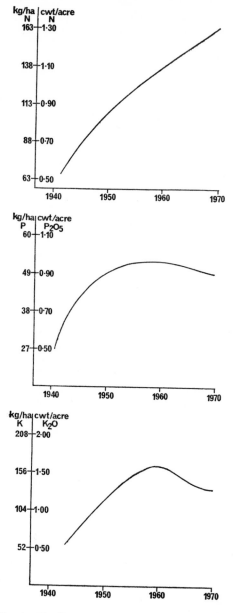

Fig. 1. Fertiliser usage for sugar beet, 1940–1970.

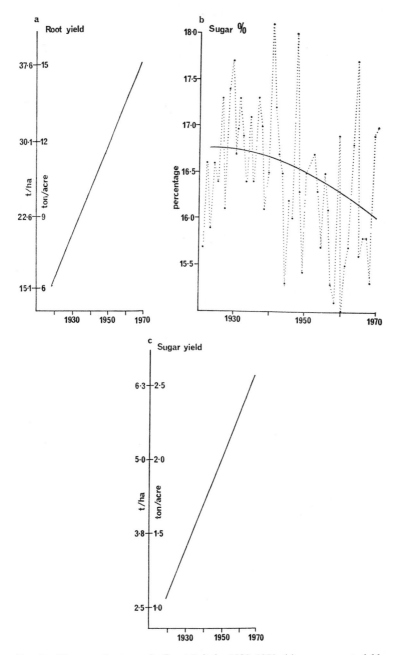

FIG. 2. The sugar-beet crop in Great Britain, 1920–1970: (a) average root yield; (b) average sugar percentage; (c) average sugar yield.

110 000 acres contains the equivalent of 0·24 cwt/acre K_2O spread over the whole 450 000 acres of sugar beet. Thus on average the crop receives $1·32 + 0·24 = 1·56$ cwt/acre K_2O in fertilisers. As about one-quarter of the crop receives farmyard manure as well, there seems to be little scope for increasing yields by supplying more potassium. Increased use of sodium would increase yields and allow considerable savings of potassium (Chapter 4).

Trends in fertiliser usage for sugar beet

Figure 1 shows the average amount of nitrogen, phosphorus and potassium used on sugar beet since the early 1940's, obtained by combining the survey data reported above and interpolating where necessary. Clearly, nitrogen dressings have increased rapidly from 0·50 cwt/acre in the early 1940's to about 1·30 cwt/acre in 1970. Phosphorus fertiliser usage nearly doubled during the period 1940–1955, but during the last fifteen years has been remarkably stable at about 0·95 cwt/acre. Usage of potassium (excluding that applied in kainit) was about 0·50 cwt/acre in 1940 but had doubled by 1950. By 1960 the average usage was 1·50 cwt/acre but has since declined to a fairly static 1·30 cwt/acre.

Trends in yields

Although the area of sugar beet grown each year is stable at 450 000 acres as a result of Government control, the amount of sugar produced from beet is increasing. This is largely due to the almost linear increase in root yield per unit area during the last thirty years (Fig. 2a). The sugar percentage of the crop fluctuates greatly from year to year due to differences in weather, but it appears to be declining slightly (Fig. 2b). However, sugar yield per unit area is increasing rapidly (Fig. 2c). The decline in sugar percentage may be partly caused by growers' preference for varieties which produce large roots with small sugar percentage ('E' types) rather than varieties with small roots and large sugar percentage ('Z' types). More likely the decline is because of the increased use of nitrogen fertiliser which increases the amount of water and impurities in the roots and so lessens the sugar percentage. In addition, during the last ten years more of the crop has been harvested and processed later than previously, which may account for part of the decrease.

Chapter 2

Nitrogen

In most sugar-beet growing regions of the world, nitrogen is the most important fertiliser element. This is because nitrogen is usually in short supply in soils under continuous cropping and few arable soils can regularly provide more than 0·5 cwt/acre N/annum. Sugar beet takes up soil nitrogen more efficiently than most crops but as it needs up to 2·0 cwt/acre N for maximum yield, some must be given in fertiliser. This usually increases yield greatly but the quality of the crop declines, particularly with large dressings. The amount of nitrogen fertiliser needed by sugar beet has been the subject of many experiments in every sugar-beet growing country.

Nitrogen in sugar beet

QUANTITY OF NITROGEN IN SUGAR-BEET CROPS AT HARVEST

As with any nutrient, the crop obtains part from applied fertiliser and part from soil reserves—in the case of nitrogen the latter is mainly in the form of decaying organic matter or unused fertiliser given for previous crops. Table 5 shows the total amount of nitrogen contained in crops of sugar beet at harvest. Nitrogen fertiliser greatly influences the amount of nitrogen in the crop—without fertiliser the crop contained between 0·2 cwt/acre when grown on soil with small reserves of nitrogen[360] and 0·8 cwt/acre when grown on relatively fertile soil.[132] With fertiliser, crops contained up to 3·2 cwt/acre.[132] Given a 'commercial' dressing (1·2 cwt/acre) an average crop will contain about 1·6 cwt/acre, about a third of which is in the roots and two-thirds in the leaves and petioles.

CONCENTRATION OF NITROGEN IN SUGAR BEET AT HARVEST

Table 6 shows the concentration of nitrogen in dried sugar beet at harvest. It is most concentrated in the leaves (which usually contain more than 3·0%) and least concentrated in the roots (0·6%); petioles contain about 1·4%. Goodman[132,133] analysed whole plants and found the average concentration was about 2%; Draycott

TABLE 5

QUANTITY OF NITROGEN IN SUGAR BEET CROPS AT HARVEST

Quantity of N (cwt/acre)					Quantity of N (kg/ha)					References
Roots	Petioles	Leaves	Total	N dressing	Roots	Petioles	Leaves	Total	N dressing	
			1·5	0·8				188	100	Adams[5]
			1·5	1·0				188	126	Draycott and Holliday[87]
			0·8–2·4	0–1·2				100–301	0–151	Goodman[132]
			1·4–3·2	0·4–1·2				176–402	50–151	Goodman[133]
0·35–0·4	0·3–0·5	0·5–0·7	1·1–1·6	0·6–1·2	44–50	38–63	63–88	138–201	75–151	Thorne and Watson[328]
0·3–0·4	0·4–0·8		0·2–0·9	0–0·8 approx	38–50	50–100		25–113	0–100	Warren and Johnston[360]
			0·7–1·2	0–0·8				88–151	0–100	Widdowson, Penny and Williams[371]
Means 0·4	0·4	0·6	1·4	0·9	50	50	75	175	113	

TABLE 6

TABLE 6

CONCENTRATION OF NITROGEN IN DRY MATTER OF SUGAR
BEET AT HARVEST

N concentration Roots Petioles Leaves (% dry matter)			N dressing (cwt/acre) (kg/ha)		References
0·5	1·1	3·0	0·8	100	Adams[5]
0·8	2·7		1·0	126	Draycott and Holliday[87]
	2·3–2·7		0–1·2	0–151	Goodman[132]
	1·0–2·0		0·4–1·2	50–151	Goodman[133]
0·6–0·8	1·2–1·8	3·1–3·5	0·6–1·2	75–151	Thorne and Watson[328]
Means 0·6	1·4	3·2	1·0	126	

and Holliday[87] analysed whole tops (lamina, petiole and crown) and
found they contained 2·7%.

AMOUNT AND CONCENTRATION OF NITROGEN IN THE SUGAR-BEET
CROP DURING THE GROWING PERIOD

Knowles et al.[203] made a study of the amount and concentration of
nitrogen in sugar beet sown at the beginning of May and given 0·6
cwt/acre N. Table 7 shows that the concentration decreased progressively throughout the season but that the amount of nitrogen in
the crop increased. Concentration in the leaves was greater than in
the roots on every occasion; early in the season the quantity of
nitrogen in the leaves was greater than in the roots but in September
it was evenly distributed between leaves and roots. More recently,

TABLE 7

NITROGEN CONCENTRATION IN SUGAR BEET DRY MATTER
AND AMOUNT IN THE CROP DURING THE GROWING PERIOD
(after Knowles et al.[203])

	N concentration Leaf Root Whole plant (% dry matter)			Amount of N in the crop Leaf Root Whole plant (cwt/acre)			Leaf Root Whole plant (kg/ha)		
May	4·8	2·7	4·4	0·13	0·01	0·14	16	1	17
June	4·6	2·4	4·2	0·42	0·06	0·48	53	8	61
July	3·0	1·0	2·0	0·57	0·19	0·76	72	24	96
August	2·0	0·8	1·2	0·43	0·36	0·79	54	45	99
September	1·9	0·8	1·1	0·46	0·47	0·93	58	59	117

Goodman[132] investigated the uptake of nitrogen by field-grown sugar beet at Rothamsted (clay with flints), Stamford (shallow limestone soil) and Holbeach (silty loam). Figure 3 shows the total amount of nitrogen taken up by the crop from May to November. The crop on the deep, fertile silt contained most nitrogen and that on the shallow limestone soil contained least.

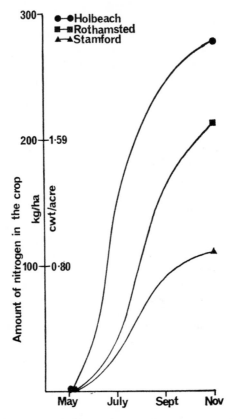

FIG. 3. Total amount of nitrogen in sugar-beet crops on three contrasting soils.[132]

Nitrogen-deficiency symptoms

STAGE OF GROWTH

Unlike deficiency symptoms of some elements, nitrogen deficiency shows on sugar-beet leaves at almost any stage of development of the plant. The seed contains sufficient nitrogen to supply the emerging cotyledon leaves but symptoms of deficiency can show on the first pair of 'true' leaves. This indicates that an external supply of nitrogen is needed to prevent development of symptoms from this stage onwards. The cotyledon leaves sometimes have symptoms of deficiency as the plant grows older.

SYMPTOMS

Nitrogen deficiency in sugar beet shows as a general yellowing of the foliage. There are no symptoms which completely characterise the deficiency as is the case, for example, with manganese, boron and magnesium. The general yellowing or chlorosis of the laminae and, to a lesser extent, of the petioles is also symptomatic of certain types of pest and disease damage.[187] In the field, nitrogen-deficient sugar beet often occur in patches, presumably due to changes from place to place in the soil nitrogen supply or to distribution of soil pathogens which damage the root system and thereby decrease uptake of nitrogen. Usually, leaves and petioles of all ages are fairly uniformly yellow when the plant is nitrogen-deficient, although the youngest leaves may be somewhat greener than the old ones. Where plants have yellow old leaves but bright green healthy young leaves, some trouble other than nitrogen deficiency should be suspected (the three most likely ones are downy mildew, virus yellows and magnesium deficiency).

When the supply of nitrogen is inadequate, sugar-beet plants appear stunted and the leaves are small with long thin petioles; the inner leaves sometimes form a rosette. Older leaves are often prostrate and the outermost ones senesce early. Symptoms are most common on sandy or gravelly soils either where the nitrogen fertiliser application has been omitted or leached. Sugar beet grown after ploughed-in straw sometimes show symptoms resulting from competition for available soil nitrogen. Excessive competition from weeds in the early stages of growth of the crop has the same effect.

Although nitrogen deficiency causes premature death of the older leaves, Loomis and Nevins[228] found that the numbers of dead leaves on plants (in a pot experiment) with and without nitrogen were similar. This was because nitrogen shortage decreased the rate of leaf initiation, the nitrogen-deficient plants having only about half the number of living leaves as the plants with an adequate nitrogen supply. More recently the same authors[255] have shown that nitrogen deficiency decreases chlorophyll concentration and the photosynthetic rate of old but not young leaves.

ANALYSIS OF DEFICIENT PLANTS

Analysis of sugar-beet leaves and petioles from experiments where different quantities of nitrogen fertiliser were tested showed that the degree of yellowing of the foliage was related to the concentration of total nitrogen in the plants.[98] Plants with dark green foliage contained from 2·6 to 3·2% N in the dry matter of the tops at harvest. Those which were distinctly yellow contained from 1·9 to 2·3% N and there was a gradation in colour and nitrogen concentration between the extremes. In California, Ulrich and Hills[347] found that petioles of sugar beet with deficiency symptoms contained 70–200 ppm nitrate nitrogen whereas petioles from plants without symptoms contained 350–35 000 ppm in dried tissue.

Effect of nitrogen fertiliser on the growth and physiology of sugar beet

Nitrogen fertiliser has pronounced effects on the growth and physiology of sugar beet, even to the extent of causing large changes in the physical and chemical characteristics of the crop at harvest. Extreme shortage of nitrogen results in deficiency symptoms described on page 13 and poor yields of roots and tops. Excess available nitrogen in the soil increases leaf growth but decreases the quality of the crop. The magnitude of some of these effects is described below.

DRY MATTER PRODUCTION

Figure 4 shows the effect of nitrogen dressings on the production of dry matter by sugar beet. The smaller of the two dressings was chosen to give near maximal production of sugar/acre whereas the larger dressing was double this amount. Although the smaller dressing gave as much sugar/acre as the larger dressing, the latter gave considerably more dry matter/acre. The extra nitrogen increased the growth of tops but gave very little more root dry matter than the smaller dressing. Figure 4 also shows that nitrogen fertiliser increased the rate of production of leaf and petiole dry matter early in the season with little effect, at that stage, on root dry matter. As the season progressed, the nitrogen fertiliser maintained the leaf and petiole production and this was reflected in increased root dry matter production.

Campbell and Viets[47] in Montana have studied the effect on the growth of sugar beet of nitrogen fertiliser in the seedbed compared with nitrogen given as a top-dressing. Figure 5 shows the effect of the two times of application of N on the production of fresh tops; it was not until mid-June that the seedbed nitrogen treatment began

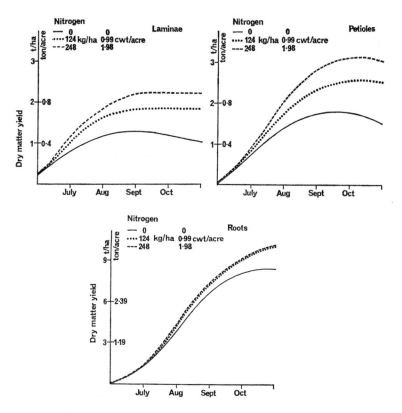

FIG. 4. Effect of applying 0, 0·99 or 1·98 cwt/acre N (0, 124 or 248 kg/ha N) on total amount of nitrogen in sugar-beet laminae, petioles and roots. Average of nine experiments.

to increase tops yield. Nitrogen as a top-dressing produced large increases in top growth regardless of the spring treatment but none increased the root yield.

DRY MATTER DISTRIBUTION

The effect of nitrogen on yield of tops and roots at Woburn (Table 8) is typical; nitrogen fertiliser invariably increases the proportion of tops which make up the ultimate total dry matter yield. In this case the increased proportion of tops from the larger dressing was simply a result of increased tops yield with no effect on root dry matter yield.

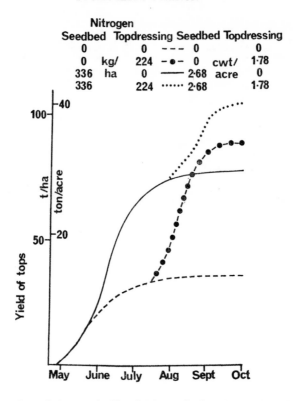

FIG. 5. Effect of nitrogen fertiliser in the seedbed and/or as a top-dressing on yield of tops. [47]

TABLE 8

EFFECT OF NITROGEN FERTILISER ON THE DISTRIBUTION OF DRY MATTER IN SUGAR BEET AT WOBURN

	Dry matter yield					
	(cwt/acre)				(t/ha)	
			N dressing			
		(cwt/acre)			(kg/ha)	
	0	0·75	1·50	0	94	188
Tops	21·3	32·9	46·8	2·67	4·13	5·87
Roots	59·7	79·8	81·1	7·49	10·02	10·18
Tops:roots	0·36	0·45	0·57	0·36	0·45	0·57

EFFECT OF NITROGEN FERTILISER ON GROWTH OF SUGAR BEET IN POT EXPERIMENTS

Milford and Watson[248] found that changes in leaf area accounted wholly for increases in dry matter produced by nitrogen fertiliser because net assimilation rate was unaffected. The fertiliser did not alter the partition of the total assimilate between roots and tops, in contrast to sugar-beet field experiments where the proportion of tops is usually greatly increased by nitrogen fertiliser. Milford and Watson showed that nitrogen fertiliser increased the fraction of the assimilate entering the roots that was used in growth at the expense of that stored as sugar. Thus plants with more nitrogen had a smaller proportion of their root dry weight as sugar because more was used in growth of the roots and not because less entered the roots.

Terry[326] grew sugar-beet plants at temperatures of 15 and 25°C with a small and a large supply of nitrogen. The first-formed leaves grew faster and larger at 15 than at 25°C. Leaves were produced, unfolded and grew faster with the larger amount of nitrogen as cells divided and expanded faster, increasing in number and size. Nitrogen concentration was also greater in the leaves and storage root at 15 than at 25°C with the larger supply. Sugar concentration was also greater at 15 than at 25°C in the leaves but not in the root. The only clearly-defined function of nitrogen in all these experiments appears to be in increasing leaf area, hence increasing dry matter production.

EFFECT OF SHORTAGE OF NITROGEN ON GROWTH

Loomis and Nevins[228] grew sugar beet in vermiculite and nutrient solution to determine the effect of nitrogen supply on growth of the leaves and storage root and on sugar percentage of the root. With a large supply of nitrogen, new leaves were initiated at a fairly constant rate. Leaf area per plant reached a maximum by mid-September and then declined due to the progressively smaller size of the new leaves. Shortage of nitrogen decreased the rate of leaf initiation, leaf area and dry matter accumulation and increased the sugar percentage. During the first nine weeks of growth the increase in sugar percentage was sufficient to compensate for the decreased weight of root, and nitrogen did not affect the sugar yield. Plants grown for a period without nitrogen did not recover completely when the supply was renewed. Leaf initiation began but the sugar yield declined. Root growth and leaf expansion were limited during the period of recovery and lower sugar yields were obtained by giving nitrogen to the deficient plants than by allowing the deficiency to continue.

Loomis and Worker[230] made a somewhat similar experiment with field-grown sugar beet in California. The crop was sown in October with a basal dressing of nitrogen and experimental nitrogen treatments applied in February, March and April of the following year so that the effects of a period of nitrogen deficiency could be measured after renewal of the nitrogen supply. Growth of storage root and of tops increased very soon after nitrogen was applied but fresh weight of tops increased at an above-normal 'compensatory' rate. However, sugar accumulated less rapidly in refertilised plants than in the roots of plants that were maintained at either continuously small or large nitrogen supply. It therefore appears that the effects of a period of nitrogen shortage cannot be entirely eliminated by applying supplemental nitrogen.

EFFECT OF NITROGEN FERTILISER ON LEAF AREA

Leaf area index (L) is a convenient way of describing the leaf area of a crop; it is the total area of leaf per plant divided by the area of land the plant occupies.[363] Campbell and Viets[47] showed how 2·7 cwt/acre N increased L throughout the growing season, particularly from early July onwards. They also showed how a top-dressing of 1·8 cwt/acre N applied in August rapidly increased L. When L is integrated over time, the expression is called the leaf area duration (D). It is a useful parameter in physiological studies and is often related to yield: for further details see Goodman,[134] who established several important relationships between D and other growth parameters for the sugar-beet crop. Goodman[132] demonstrated that D was greatly influenced by nitrogen supply from soil and fertiliser. On average of measurements made at three locations, D for plants grown without nitrogen fertiliser was 23 weeks, and with about 1·0 cwt/acre N fertiliser D was 31 weeks. In later experiments, corresponding values were 35 (without N) and 43 weeks (1·8 cwt/acre N) at Rothamsted and 16 (0 cwt/acre N) and 36 weeks (1·8 cwt/acre N) at Broom's Barn. Thus the magnitude of the effect of nitrogen fertiliser is variable but always increases D.

PHOTOSYNTHETIC EFFICIENCY

The efficiency of the crop canopy to synthesise dry matter is measured by the net assimilation rate (E).[363] Goodman reported that nitrogen slightly decreased E in both his early experiments[132] and in later ones[133,134] but the differences were not statistically significant. Campbell and Viets[47] in California found that E was not affected consistently by nitrogen fertiliser. It is questionable whether any of these small effects are meaningful for E is difficult to determine accurately in field experiments.

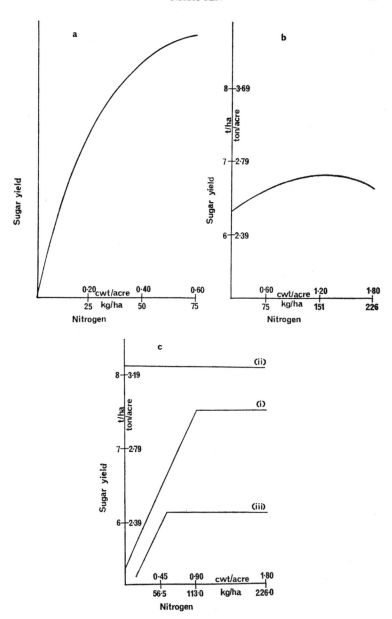

FIG. 6. Nitrogen application and sugar yield: (a) exponential form used by Crowther and Yates;[71] (b) parabolic form proposed by Boyd;[34] (c) 'two straight line' relationship proposed by Boyd et al.[38]

Effect of nitrogen fertiliser on sugar yield

The effect of nitrogen fertiliser on yield of sugar has been measured in many field experiments in this country and abroad. On most soils in all parts of the world increasing increments of nitrogen fertiliser at first rapidly increase the amount of sugar produced by the crop. Eventually, further increments increase yield little—indeed there are many reports that excessive dressings decrease yield. Taking into account the large number of experiments made, it is rather surprising that there have been few investigations of the precise nature of the nitrogen/sugar yield response curve.

Crowther and Yates[71] assumed and used an exponential curve (Fig. 6a) to estimate the nitrogen requirement of sugar beet. The form adopted was similar to that used by Mitscherlich, with no provision for decreases in yield with large dressings of fertiliser. However, in 1961 Boyd[34] proposed a parabolic curve (Fig. 6b), suggesting a clearly defined optimum dressing and that, on average, more than 1·20 cwt/acre N decreased sugar yield considerably.

TABLE 9

EFFECT OF NITROGEN FERTILISER ON SUGAR YIELD

			Sugar yield		
N dressing	California[376]	Canada[16]	Great Britain[38]	Israel[141]	Mean
(cwt/acre)			(cwt/acre)		
0	59·4	61·4	45·1	45·4	52·9
0·35	67·1	64·2	50·0	53·3	58·7
0·70	74·7	65·0	54·1	59·7	63·4
1·05	78·8	61·9	54·8	62·1	64·4
1·40	81·2	61·3	54·1	62·1	64·7
1·75	81·0	62·3	54·3	61·3	64·5
2·10	79·4	60·2	—	—	—
2·45	78·8	—	—	—	—
(kg/ha)			(t/ha)		
0	7·46	7·71	5·67	5·70	6·64
45	8·43	8·06	6·28	6·70	7·37
90	9·39	8·16	6·79	7·50	7·96
135	9·90	7·78	6·88	7·80	8·09
180	10·20	7·70	6·80	7·80	8·13
225	10·18	7·83	6·82	7·70	8·10
270	9·97	7·56	—	—	—
315	9·90	—	—	—	—

Some values interpolated.

During the last ten years 170 further experiments have been made, many testing up to 1·80 cwt/acre N. As a result, Boyd et al.[38] have now proposed that the relationship takes the form of two straight lines; on most fields, nitrogen fertiliser initially increased yield greatly until a point was reached [Fig. 6c(i)] where more nitrogen had little effect on sugar yield. On some fields, nitrogen fertiliser had no effect throughout the range (ii) which leads to the conclusion that the supply from the soil was adequate for maximum sugar accumulation. On other fields there was a small response to fertiliser (iii).

Sugar yield responses to nitrogen fertiliser in experiments from several other countries where four or more amounts of nitrogen have been tested are shown in Table 9. They lead to the conclusion that response follows the form proposed by Boyd et al.,[38] illustrated in Figure 6c. Nitrogen given for sugar beet on some fields increased yield considerably and a large dressing was needed, but excess had little effect. On others even a small amount of nitrogen did not increase yield, but neither did it decrease it.

Factors which may affect the amount of fertiliser needed

SOIL TYPE

Mineral soils
Shepperd[309] investigated the supposition of many sand land farmers in Norfolk that sugar beet needed more nitrogen than on heavier soils. However, as on many soils, 0·90 cwt/acre N was found to give maximum yield of roots; more than 0·60 cwt/acre decreased sugar percentage, particularly in a dry season. Bland[24] made a similar investigation of the nitrogen fertiliser requirement of the crop on clay soil, also in Norfolk. He, too, concluded that 0·90 cwt/acre gave maximum yield of roots and sugar but that each successive further addition of 0·20 cwt/acre gave an extra 1 ton/acre tops.

Boyd et al.[32] in a large-scale investigation of fertiliser requirement of sugar beet in Great Britain found little difference in nitrogen requirement on 300 fields in different parts of the country; organic soils were the only exceptions, needing less nitrogen. The value of these experiments was somewhat limited because of the narrow range of amounts of nitrogen tested—the largest dressing was only 0·8 cwt/acre.

Adams[8] tested up to 1·8 cwt/acre N on 49 fields and found that the difference in response between soil types was small in comparison to the difference in response between fields on the same soil type. The only soils which he was able to pick out where sugar beet responded differently from the average were calcareous clays of glacial origin (Hanslope and Stretham Series):

| | Cwt/acre N | | |
	0·6	1·2	1·8
Sugar (cwt/acre)			
Hanslope and Stretham Series	51·6	54·0	54·6
All soils	52·8	53·4	51·7

The average requirement was about 0·90 cwt/acre, but on the Hanslope and Stretham series the sugar beet needed more than 1·20 cwt/acre for maximum yield. Draycott[84] confirmed this greater need of sugar beet on the calcareous clays of SE England in later experiments.

Both Boyd et al.[32] and Adams[8] based their conclusions on the assumption that there was a quadratic (or parabolic) relationship between sugar yield and nitrogen dressing; this is now known to be incorrect (see page 21). Boyd et al.[38] have therefore recently reviewed results from 170 experiments testing nitrogen in Great Britain between 1957 and 1966 (including those of Adams[8] and Draycott[84]). Although there appeared to be much variation in response from field to field, they pointed out that only part of the apparent difference came from real differences in crop requirement—responses also differed because of experimental error, which often accounted for much of, and sometimes all, the between-site variation in response. There were substantial between-site differences in response to amounts of N up to 0·9 cwt/acre but attempts to explain them in terms of soil (or any other factor) had little success. Differences from field to field in response between 0·90 and 1·80 cwt/acre N were usually no greater than could be expected from experimental error alone. This suggested that sugar beet rarely needed more than 0·90 cwt/acre. The Hanslope and Stretham series were the only soils where sugar beet consistently needed slightly more than 0·90 cwt/acre.

The conclusion from these experiments testing nitrogen requirement of sugar beet on various mineral soils is that neither textural nor pedological classification is of much value in predicting requirement. It is remarkable that few crops need more than 0·90 cwt/acre N for maximum sugar yield when total nitrogen uptake by the crop varies greatly between soil types. For example, Goodman[132] showed that fertiliser recovery was $5\frac{1}{2}$ times that applied on a deep silty loam to $1\frac{1}{2}$ times that applied on a shallow loam over limestone—totals respectively of 4·0 cwt/acre and 1·8 cwt/acre N. Clearly, many factors other than soil type and N supplied in fertiliser determine yield and hence nitrogen uptake.

Organic soils

Boyd *et al.*[32] described experiments in the years 1934–49 with sugar beet, including 32 on organic soils. Response to nitrogen was less than on mineral soils. They suggested that sugar beet needed little nitrogen fertiliser because the soil could usually supply sufficient for maximum yield. Tinker[333] described experiments on deep peat soils in 1963–65 to determine the amount of nitrogen needed by sugar beet. The element only increased yield significantly on one out of 18 fields and he concluded that an 'insurance' dressing of 0·4 cwt/acre N was all that was needed. Draycott[94] described 18 further experiments in 1966–69 on peaty mineral and organic mineral soils. Nitrogen had a larger average effect than in Tinker's experiments, 0·6 cwt/acre N increasing root yield by 1 ton/acre, and 1·20 cwt/acre by a further $\frac{1}{2}$ ton/acre. The dressing for maximum sugar yield was 0·60 cwt/acre.

Draycott and Durrant[101] examined the response to nitrogen fertiliser in the 36 experiments 1963–69[333,94] and in a further 16 ADAS experiments on organic soils in relation to the loss on ignition of the soil. As the loss on ignition declined the response to nitrogen fertiliser increased:

Loss on ignition (%)	Increase in sugar yield (cwt/acre)
>71	+0·3
61–70	+1·4
51–60	+1·1
36–50	−0·3
26–35	+3·6
14–25	+6·1

Grouping the soils by total nitrogen concentration generally predicted the response to nitrogen fertiliser, and sugar beet grown only on soils in the range 0·50–0·75% N responded well to nitrogen fertiliser. With more than 0·75% N the response in sugar yield was worth less than the cost of the fertiliser.

LENGTH OF GROWING PERIOD

The length of growing period of the sugar-beet crop varies greatly from country to country. Even within Great Britain, by combinations of early or late sowing and early or late harvesting the crop is growing from less than 150 days to more than 300 days. This raises the question whether length of growing period (or time of sowing or harvesting) affects the fertiliser requirement of the crop. Holmes and Adams[183] reported two series of experiments to investigate this—one in England and one in Scotland. Table 10 shows the average

TABLE 10

EFFECT OF TIME OF SOWING AND HARVESTING ON RESPONSE TO FERTILISER

(after Holmes and Adams[183])

	Sugar yield response					
	(cwt/acre)			(t/ha)		
	Early	Medium	Late	Early	Medium	Late
		Sowing date			Sowing date	
England (mean of four experiments)	+7·8	+7·6	+5·9	+0·98	+0·95	+0·74
Scotland (mean of two experiments)	—	+5·6	+5·1	—	+0·70	+0·64
	Early	Medium	Late	Early	Medium	Late
		Harvesting date			Harvesting date	
England (mean of four experiments)	+6·5	+6·5	+8·2	+0·82	+0·82	+1·03
Scotland (mean of five experiments)	+3·9	+3·3	+2·7	+0·49	+0·41	+0·34

effect of time of sowing and harvesting on response to compound fertiliser containing nitrogen. On average, response tended to decrease with late sowing and with early harvesting, although the results were not very consistent. The tentative conclusion is that response to fertiliser increases with increasing length of growing period. This is largely to be expected, for yield increases greatly.[189A] There was little evidence, however, that more fertiliser was needed as the growing period was extended and more, large and refined experiments are needed to investigate this in detail. Jorritsma[198] reviewing responses by sugar beet to nitrogen in Holland also suggested that increasing the length of the growing period did increase the amount of nitrogen fertiliser required.

Last and Tinker[217] and others (see page 31ff) have shown that the concentration of nitrate in sugar-beet tissue is a measure of the nitrogen status of the crop; also that the concentration decreases greatly with time. Ulrich[341] showed that sugar is most concentrated in the roots when the crop is nitrogen-deficient (contains less than 1 000 ppm NO_3^--N in dry petiole tissue) for about 11 weeks before harvest. This has also been confirmed in England.[217] As nitrogen fertiliser applications increase the NO_3^--N concentration, this is indicative that the amount of fertiliser needed may well vary with length of growing period.

Schmehl et al.[303] tested the effect of 0, 0·36 and 1·08 cwt/acre N given for sugar beet planted on 7th April or 3rd May in Colorado. Table 11 shows that 0·36 cwt/acre N increased sugar yield with early

TABLE 11

EFFECT OF NITROGEN FERTILISER AND SOWING DATE ON SUGAR YIELD IN COLORADO, USA
(after Schmehl et al.[303])

N dressing		Sugar yield			
		(cwt/acre)		(t/ha)	
(cwt/acre)	(kg/ha)		Sowing date		
		7 April	3 May	7 April	3 May
0	0	60·4	61·0	7·58	7·66
0·36	45	65·8	58·4	8·26	7·33
1·08	135	61·0	55·0	7·66	6·90

sowing but the same amount applied for the late-sown crop gave no increase: 1·08 cwt/acre N even decreased yield when given to the late-sown crop.

Boyd[37] compared the mean yields in 170 experiments from sugar beet given 0, 0·6 and 1·2 cwt/acre N and harvested between September and December with the yields from 17 experiments harvested after 20th November (Table 12). The response with each amount of nitrogen was greater for the late-harvested crops than for all crops taken together. This suggests that the response to nitrogen fertiliser increases if harvesting is delayed.

The few reports available are generally in agreement that the earlier the crop is sown the larger the amount of nitrogen fertiliser needed to give maximum yield. With late sowing the nitrogen has less beneficial effect than for early-sown sugar beet and there is some evidence that it is harmful. The effect of harvesting date on nitrogen requirement has not been fully investigated but it is tentatively suggested that the later the harvest the larger the response to nitrogen, with the implication that the requirement of fertiliser nitrogen by the crop may also be greater.

TABLE 12

EFFECT OF TIME OF HARVESTING AND AMOUNT OF NITROGEN
FERTILISER ON SUGAR YIELD
(after Boyd[37])

	No. of *experiments*	*Sugar yield*					
		(cwt/acre)			*(t/ha)*		
				N dressing			
		(cwt/acre)			*(kg/ha)*		
		0	0·6	1·2	0	75	150
Late harvested							
(after 20 November)	17	40·6	52·1	55·4	5·10	6·54	6·95
All experiments	170	46·3	55·0	56·4	5·81	6·90	7·08

WINTER RAINFALL

Boyd et al.[32] showed a close association between winter rainfall and
nitrogen response ($r = +0·71$) for the period 1933–51. It alone
accounted for about half the variance between responses. One inch
above (or below) the mean rainfall per month for the five months
November to March was sufficient to increase (or decrease) the
nitrogen response by $2·54 \pm 0·48$ cwt/acre sugar. The most obvious
explanation of this effect is that the rain leaches out the nitrate
nitrogen from the soil. More recently, however, Boyd et al.[38]
investigated the response to nitrogen in 1957–1966 and found that
the rainfall in the previous winter did not account for a significant
part of the variation in response.

Correlations of this sort made on experiments from field to field
are always influenced by secondary effects such as previous cropping
and manuring. It is shown in Chapter 9 that the previous crop, the
amount of nitrogen given to it, and the residue of fertiliser and plant
nitrogen left in the soil greatly affect the amount of nitrogen needed
by sugar beet for maximum sugar yield. If any correction is made for
rainfall in the winter prior to sugar beet, then it appears that this will
only be of the order of $\pm 0·30$ cwt/acre, and then only in exceptionally
dry or wet winters.

SUGAR-BEET VARIETY

Varieties of sugar beet differ greatly in growth habit, some producing
large yields of roots of relatively small sugar percentage (the 'E'
types) whilst others produce smaller root yields with large sugar
percentage ('Z' types). Yields of leaf, petiole and crown also differ
greatly between varieties. With such variations in dry matter production, it seems possible that different amounts of nitrogen fertiliser
may be needed for maximum yield.

Goodman[133] tested 0·4, 0·8 and 1·2 cwt/acre N on Sharpe's Klein 'E' and Klein 'Z' (Kleinwanzleben Zucker West Zone, a diploid variety) at Broom's Barn, Rothamsted and on an organic soil. The two varieties did not differ consistently in uptake of nitrogen or in nitrogen concentration in the dry matter at the three centres. There was a significant interaction between nitrogen and variety in total dry yield (the 'Z' variety needed most nitrogen to give maximum yield) but no interaction between them in sugar yield.

In Israel, Gutstein[141] found a large significant interaction in sugar yield between farmyard manure, nitrogen and the two varieties Kleinwanzleben Cerco Poly (an anisoploid) and the diploid Zwaanesse III. Cerco Poly gave maximum yield with nitrogen fertiliser alone whereas Zwaanesse III needed manure as well. The author concluded that the Cerco Poly had a more effective root system which was able to take up other major and minor nutrients required for maximum growth from the soil reserve, whereas Zwaanesse III needed a supply from manure. These findings need further confirmation elsewhere.

Desprez[78] reported different responses to nitrogen by diploid and polyploid varieties in France; polyploids needed slightly more nitrogen than diploids. Jorritsma[198] in Holland found that the old diploid 'N' types ('N' types are intermediate between 'E' and 'Z') such as Kuhn P and Hilleshög-standard needed more nitrogen than Kleinwanzleben E (another diploid but with large tops), but polyploid varieties had a greater nitrogen requirement than the Kleinwanzleben E. Kuhn P as well as the polyploids appeared to be less susceptible to over-dosage than Kleinwanzleben 'E'.[199]

Hills et al.[177] in experiments with American varieties in California grew each [two 'Z' types SL 824 (US 35/2) and SL 828 and two 'N' types US 33 and US 22/3] with 0, 0·7, 1·4 or 2·1 cwt/acre N. There was an interaction between varieties and nitrogen in both root and sugar yield for US 35/2 responded to more nitrogen than US 22/3 and the authors suggested that this was probably because the 'Z' type had a less efficient root system than the 'N' type.

In conclusion, results from different parts of the world suggest that polyploid varieties may need more nitrogen for maximum yield than diploids, but further investigations are needed. There is also some evidence that the order of requirement by the different sugar types is 'Z' > 'N' > 'E' but here again, no detailed experiments have been made. From the experiments available it is not certain that there are differences in fertiliser requirement. If there are differences in the optimal nitrogen dressings between varieties they appear to be small, probably of the order 0·2–0·3 cwt/acre for the extremes of varieties or types.

Interactions between nitrogen and other fertiliser elements

PHOSPHORUS

Boyd et al.[32] described a comprehensive series of over 300 experiments in Great Britain between 1934 and 1949 testing nitrogen and phosphorus in factorial combination. Nitrogen increased sugar yield (+3·73 cwt/acre) more than phosphorus (+1·55 cwt/acre) but there was a small positive interaction between them (+0·26 cwt/acre). Adams[8] made 41 similar experiments more recently (1957–60) and found no interaction between nitrogen and phosphorus in sugar yield, probably because the available phosphorus in the experiments of Adams was greater than in those of Boyd et al. due to increased use of fertiliser. Jónsson[196] in Sweden in a thorough investigation on soils of various phosphorus levels also found no interaction between nitrogen and phosphorus.

POTASSIUM

In experiments reported by Boyd et al.[32] the N × K interaction was the largest of the interactions between the three major nutrients nitrogen, phosphorus and potassium. The average of more than 300 fields was +0·77 cwt/acre sugar. However, Adams[8] found this positive interaction was rarely significant. Tinker,[329] testing nitrogen and potassium, found that when the nitrogen dressing was increased from 0·60 to 1·20 cwt/acre without potassium, sugar yield increased by 0·6 cwt/acre, but with potassium the yield increased by 1·4 cwt/acre, indicating a large positive interaction. Gallagher[124] also found a significant positive interaction between nitrogen and potassium in yields of sugar from experiments in Eire. Table 13 illustrates that 0·53 cwt/acre N increased yields of sugar by 0·9, 4·3 and 4·4 cwt/acre when 0, 1·29 and 2·58 cwt/acre K_2O was applied, whereas in the same experiments the mean effect of the N alone was to increase yield by only 3·2 cwt/acre.

Widdowson and Penny[370] in a rotation experiment at Woburn found a large, positive and significant interaction between nitrogen and potassium in yield of sugar-beet roots (+6·0 cwt/acre dry matter) but a small negative interaction between them in yield of tops (−1·0 cwt/acre). However, in Holland, Jorritsma[198] found no interaction between nitrogen and potassium (or sodium). He concluded that this was probably because of the large concentration of potassium in the silty Dutch soils; also because of the presence of much sodium in the soil.

Nitrogen and potassium not only affect the yield of sugar but also the amount of these two elements in the roots affects the processing quality. Several experiments have been made to find out how the ratio of $N:K_2O$ given in the fertiliser affects yield and quality of the crop. Von Müller *et al.*[352] and Heistermann[166] in Germany found that with a wide $N:K_2O$ ratio, *e.g.* $1:2$ or $1:3$, the crop produced a large yield of roots with a large sugar percentage. Also with a wide

TABLE 13
EFFECT OF NITROGEN AND POTASSIUM ON SUGAR YIELD
(after Gallagher[124])

| | | \(cwt/acre\) | | | *Sugar yield* *N dressing* | | \(t/ha\) | |
| | | *(cwt/acre)* | | | | | *(t/ha)* | |
		0	0·54	1·08		0	68	136
K dressing	0	38·0	38·9	37·1		0	4·77	4·88	4·66
(cwt/acre	1·30	41·1	45·4	43·6	(kg/ha K)	136	5·16	5·70	5·47
K_2O)	2·60	43·0	47·4	45·8		272	5·40	5·95	5·75

$N:K_2O$ ratio the concentration of harmful nitrogen in the roots declined earlier in the autumn than when the ratio was narrow.

SODIUM

Adams[6] first demonstrated that the interaction between nitrogen and sodium is large, positive and of commercial importance (Table 14). Tinker[329] reporting results of a series of experiments specially

TABLE 14
EFFECT OF NITROGEN AND SODIUM ON SUGAR YIELD
(after Adams[6])

| | | \(cwt/acre\) | | *Sugar yield* *N dressing* | | \(t/ha\) | |
| | | *(cwt/acre)* | | | | *(kg/ha)* | |
		0·6	1·2			75	150
Na dressing (cwt/acre NaCl)	0	56·6	56·2		0	7·10	7·05
	2	58·8	60·4	(kg/ha Na)	100	7·38	7·58
	4	59·7	61·4		200	7·49	7·71

designed to investigate interactions in sugar-beet yields showed that 0·60 cwt/acre N was the optimal dressing in the absence of sodium but 1·20 cwt/acre N was profitable where agricultural salt was also applied.

MAGNESIUM

Tinker[330] made experiments on fields where sugar-beet crops were expected to show signs of magnesium deficiency on the leaves. Nitrogen and magnesium fertiliser were tested separately and together, and it was noticed that both decreased symptoms of deficiency. There was also a small negative interaction in yield of sugar—the response to magnesium decreased from 4·8 cwt/acre with a small dressing of nitrogen to 3·1 cwt/acre with a large dressing of nitrogen. The interaction was so small compared with the effect on symptoms that he concluded that the use of large nitrogen dressings which mask the symptoms may prevent magnesium fertiliser being applied where it would increase sugar yield.

PHOSPHORUS AND POTASSIUM

On average of many experiments, Boyd et al.[32] showed that the second order interaction N × P × K was extremely small (0·08 cwt/acre sugar) but the response in sugar yield to nitrogen alone was 2·78 cwt/acre. Response to nitrogen was almost doubled (4·84 cwt/acre) when phosphorus and potassium were also given, largely due to the first order interactions N × P and N × K. The combined effect of the three nutrients was given by the sum of the three main effects plus the second order interaction (N × P × K): 3·73 + 1·55 + 2·13 + 0·08 = 7·49 cwt/acre, which was 19% of the mean yield. (The first order interactions do not enter into this expression.)

Effect of nitrogen fertiliser on seedling emergence

The interval of time between nitrogen fertiliser application, the amount applied, the form and the method of application affect the number of seedlings which emerge. Depending on whether the crop is sown to a stand or hand-singled, this may or may not affect the final plant stand or the eventual yield. As the area of sugar beet grown without hand labour is increasing rapidly it is important that fertiliser toxicities do not affect establishment of a satisfactory plant population. Thielebein[327] in Germany found that soil type and amount of moisture in the soil were the two most important factors governing the amount of damage to the germinating seed and the seedling; ammonium salts were particularly damaging. Soil conductivity measurements were only reliable for diagnosing damage due to salt concentrations when the kind of fertiliser applied was known.

Adams[7] applied various amounts of ammonium sulphate for sugar beet and counted the number of seedlings which emerged. Table 15 shows that the emergence was decreased by 50% when

TABLE 15

EFFECT OF NITROGEN AS AMMONIUM SULPHATE ON SEEDLING
EMERGENCE
(after Adams[7])

	N dressing			
	(cwt/acre)			
	0·0	0·6	1·2	1·8
Number of seedlings (per yard)	17·6	16·3	14·8	12·3
	(kg/ha)			
	0	75	150	225
(per metre)	19·3	17·8	16·2	13·5

1·8 cwt/acre N was given as ammonium sulphate. Sawahata and
Takase[302] confirmed these findings in Japan and discovered that
urea was less damaging than ammonium sulphate; also ureaform,
which ammonificates slowly, was even less damaging than urea.
Table 16 shows that sowing sugar beet on the day following nitrogen

TABLE 16

EFFECT OF NITROGEN FERTILISER APPLIED ON FOUR DATES ON
SEEDLINGS EMERGING. SOWING DATE: 31 MARCH

0·8 cwt/acre (100 kg/ha) N as 'Nitro-Chalk' applied on:	Number of seedlings	
	(per yard)	(per metre)
24 February	20	22
9 March	21	23
25 March	18	20
30 March	16	18

fertiliser application decreased the number of seedlings slightly.
The experiment was made at Broom's Barn using 0·8 cwt/acre N as
'Nitro-Chalk'. The results suggested that the 'safe' interval between
application and sowing is one to two weeks, a factor which needs
careful consideration when sowing to a stand.

Prediction of nitrogen fertiliser requirements by plant analysis

Cooke[66] has discussed the background to plant analysis as a guide
to fertiliser need and illustrated its use in planning fertiliser applications, mostly to perennial crops. Much research has been done to

find a method of plant analysis for predicting the nitrogen requirement of sugar beet. In California, Ulrich and his co-workers have
made the greatest contribution in this field. They defined a method
of analysis which has been used on a practical scale in USA and is
being tested in many other countries.

LEAF AND PETIOLE NITRATE

Leaves and petioles of healthy sugar-beet plants contain large
amounts of nitrogen in the form of the nitrate ion (NO_3^-) [251,380]
and Ulrich and others[340,341,344] in America, Sorensen[319,320] in
Denmark, and White[366] in England have measured the concentration of nitrate to decide whether it is diagnostic of the nitrogen status
of the crop.

Ulrich *et al.*[344] and Brown[43] recommended using the youngest
fully expanded leaf; it is the only one that can be defined easily and

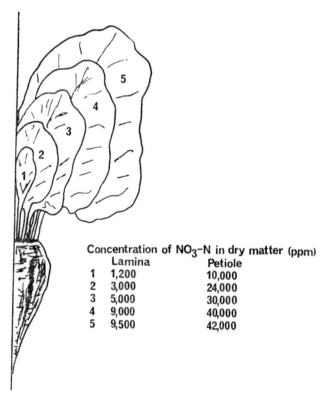

	Concentration of NO_3-N in dry matter (ppm)	
	Lamina	Petiole
1	1,200	10,000
2	3,000	24,000
3	5,000	30,000
4	9,000	40,000
5	9,500	42,000

FIG. 7.   Concentration of nitrate–nitrogen in dry matter of laminae and petioles.

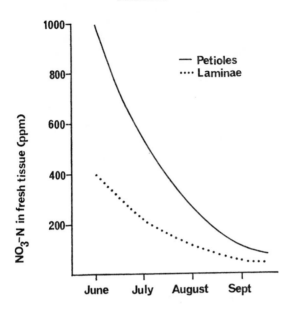

FIG. 8. Change with age in the nitrate–nitrogen concentration of the youngest fully-expanded lamina and petiole.

most workers have used it. Individual leaves on each plant differ greatly in nitrate concentration.[342,43] Figure 7 shows that generally the concentration increases considerably from the youngest to the oldest leaves.[202] The age of the plant also greatly affects the nitrate concentration of the youngest fully expanded leaf and petiole, the concentration decreasing with age as shown in Fig. 8.

ULRICH'S 'CRITICAL VALUE'

Ulrich[341,346] introduced the concept of a 'critical concentration' for sugar beet below which a plant is considered nitrogen-deficient. He suggested 1 000 ppm of nitrate nitrogen in the dry matter (100 ppm in fresh matter) of petioles as this critical concentration. When the petiole nitrate concentration was greater, yields were no greater *at the time of sampling.* However, even when 1 000 ppm nitrate nitrogen early in the season may be the critical concentration for yield at that time, this does not necessarily imply maximum yield at harvest. Ulrich[341] considered that nitrate concentration should be less than the critical value for 11 weeks before harvest, to improve sugar accumulation.

EXPERIMENTS IN ENGLAND ON LEAF AND PETIOLE NITRATE

As the amount of nitrogen fertiliser needed by sugar beet for maximum sugar yield varies from field to field, Last and Tinker[217] suggested that a small application of nitrogen might be made in the seedbed and the rest (based on plant analysis) given in June. The petiole nitrate concentration decreased rapidly with time, from about 1 000 ppm (in wet tissue) in early June to less than 100 ppm in early September. On average, petiole nitrate concentrations of about 800 ppm in June were associated with the largest sugar yields, but they considered that the method was not accurate enough to predict nitrogen top-dressing requirement of sugar beet on individual fields. Although the nitrate concentration was greatly increased by nitrogen fertiliser and seemed a good indication of the immediate nitrogen status of the crop, both the effect of age of plants and of nitrogen dressings differed between experiments. This, together with the possibility of a varying supply of soil nitrogen during the rest of the season, made it difficult to estimate with any accuracy the amount of additional nitrogen fertiliser needed.

USING THE PETIOLE NITRATE TEST

Albasal et al.[10] in Israel found a close linear relationship ($r = 0.81$) between yield of roots of autumn-sown sugar beet and petiole nitrate concentration. Nitrate concentration reached its maximum value in December, about three months after seedling emergence. They stressed that the crop should not be allowed to become short of nitrogen during this period otherwise loss of yield results. Hale and Miller[151] in California tested the technique of petiole analysis to decide whether it was suitable for predicting optimum harvest date. They found that the nitrate concentration in the petiole was not sufficiently correlated with sugar yield for it to be useful in this way. Also in California and using the same technique, Hills et al.[178] made experiments to determine how long sugar beet should be deficient in nitrogen (i.e. <1 000 ppm nitrate–nitrogen in the petiole) prior to harvest to obtain maximum sugar percentage. Analysis of the petioles did not give the answer. Even with apparent mid-season nitrogen deficiency, there was little effect on the rate of root growth and they concluded that more research was needed into the value of the test in commercial practice.

Predicting the amount of nitrogen fertiliser needed for sugar beet by soil analysis

The average amount of nitrogen fertiliser needed by sugar beet in Great Britain has been established by many experiments as 1·0 cwt/acre. Sugar beet on most fields yield best with this dressing but about 15% of crops need appreciably less or more. An average crop of sugar beet contains about 1·6 cwt/acre N. If only 1·0 cwt/acre N has been applied to the crop, at least 0·60 cwt/acre N will have been obtained from soil reserves. As the recovery of nitrogen fertiliser is only 50–70%, the soil must contribute about 1·0 cwt/acre N. This is possible because some of the organic nitrogen in soil is mineralised and becomes available to crops each year. If an analytical method could predict this amount, the optimum quantity of nitrogen fertiliser needed could be adjusted for individual fields.

The amount of potentially available nitrogen in soil is generally

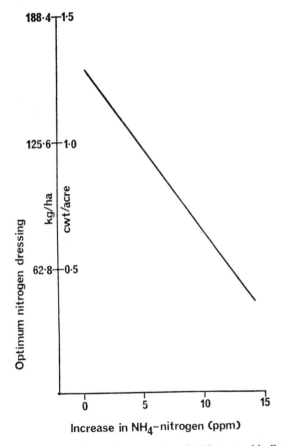

FIG. 9. Amount of ammonium-nitrogen mineralised in anaerobically incubated soil and the optimum nitrogen dressing for sugar beet.[218]

determined by measuring the mineral nitrogen before and after incubation of soil under standardised conditions. In 1961–65, soils from 65 fields throughout Great Britain were analysed in this way and field experiments made to measure the optimum nitrogen dressing for sugar beet.[218] The soils were sampled during autumn and spring and incubated either fresh or air-dry, all aerobically and some anaerobically.

The amounts of nitrogen mineralised during incubation generally did not predict with accuracy the amount of nitrogen fertiliser needed for maximum yield. The amount of mineral nitrogen in air-dry, spring-sampled soil was moderately well related to the optimum nitrogen dressing and to the increase in sugar yield from nitrogen fertiliser. The nitrogen mineralised during anaerobic incubation was best related to the optimum nitrogen dressing (Fig. 9).

Conclusions

The most difficult problem in sugar-beet nutrition is deciding the amount of nitrogen fertiliser needed. Nearly all crops need some because it is in short supply in arable soils, but too little causes serious loss of yield and too much depresses sugar percentage, purity of the juice, and the amount of white sugar extracted in the factory. However, sugar yield is increased linearly by increasing amounts of fertiliser up to 0·90 cwt/acre N on most fields, but more neither increases nor decreases it much. Thus where growers are not paid any premium for quality, it is in their interests to give a little more than 0·90 cwt/acre N to ensure that the crop has enough.

Methods of soil analysis are being investigated which may eventually predict the amount of nitrogen fertiliser needed, but none is suitable at present for use in farming practice. Although crops often contain up to 2·0 cwt/acre N, they rarely respond to more than half this amount given in fertiliser. Thus the soil supply of nitrogen greatly affects the amount which needs to be given in fertiliser. Determining the amount available from soil is difficult, partly because of the great depth from which sugar beet extracts nitrogen (about 6 ft) but even more so because soil organic matter continually releases available nitrogen at rates and in amounts which are hard to forecast. Incubation methods of soil analysis may be able to predict the nitrogen potentially available to the crop and, hence, the amount of fertiliser needed.

Another approach is to analyse the crop and apply additional fertiliser where the plants are deficient. Determination of the nitrate-nitrogen concentration in petioles or laminae is a guide to the nitrogen status of the crop and in some countries the technique has been used in farming practice. However, recent work has shown that a shortage of nitrogen in field-grown sugar beet is only damaging during the early stages of growth; most crops can obtain enough from soil during late summer and autumn to give maximum sugar yield. Giving more nitrogen from mid-season onwards is more likely to depress sugar yield than increase it. Greenhouse and field experiments have also shown that nitrogen given after the critical early stage does not make up for early shortage and depresses sugar percentage and juice purity more than nitrogen given at sowing.

A satisfactory method of chemical analysis which will predict nitrogen fertiliser requirement reliably is urgently needed. However, experiments in many countries have shown that 1·0 cwt/acre N given in the seedbed of spring-sown crops places the sugar yield on the

TABLE 17

SUMMARY OF OPTIMUM NITROGEN FERTILISER DRESSINGS IN RELATION TO SOIL, ORGANIC MANURING AND PREVIOUS CROPPING

| | N dressing | |
	(cwt/acre)	(kg/ha)
SOIL		
Mineral		
Sands, loams and silts	1·00	125
Calcareous glacial clays	1·20	150
Organic		
Loss on ignition >50%	0·40	50
Loss on ignition 10–50%	0·60	75
Compacted soils	1·20	150
ORGANIC MANURES		
Without organic manure	1·00	125
With farmyard manure	0·75	95
With slurries and poultry droppings	0·50	65
PREVIOUS CROPPING		
Two or more cereals	1·00	125
One or two year cut ley		
One cereal	0·75	95
Potatoes without farmyard manure		
One or two year grazed ley		
Permanent grassland		
Beans, peas	0·50	65
Potatoes with farmyard manure		
Clover, lucerne		
Green manure		
Vegetables		

flat part of the nitrogen response graph rather than the steeply rising portion. Also, Table 17 shows the important factors which can be taken into account when deciding the dressing. Sugar beet on organic soils invariably needs less fertiliser than average but on compacted soils which impede root penetration the crop needs more (page 165). Sugar beet grown with organic manure needs considerably less fertiliser than crops without it (page 133). Previous cropping also greatly affects the optimum dressing and in Table 17 crops and cropping systems which leave similar residues have been grouped together (*see* page 148 and ref.183A). Other factors which affect nitrogen requirement, such as irrigation, plant density and pests and diseases, are discussed in later chapters.

Chapter 3

Phosphorus and Sulphur

Much is known about the phosphate requirement of sugar beet but relatively little about the other important nutrient anion, sulphate. The crop takes similar quantities of phosphorus and sulphur from the soil but attention has been directed mainly towards the phosphorus nutrition of this and other crops because it is often in short supply, particularly in soils not previously cultivated, whereas enough sulphur is usually deposited in rain to satisfy crop requirements. However, the position may be changing slowly for in many sugar-beet producing countries the crop gives only small increases in yield from phosphorus fertiliser. This is because regular applications have increased soil supplies of available phosphorus so that they are now nearly adequate for maximum yield without fresh additions every year. In future, fertiliser will only be needed periodically to maintain a satisfactory concentration of available phosphorus in the soil. In countries where atmospheric pollution is less than in Great Britain, rain deposits much less sulphur and some crops show deficiencies. Now that ammonium sulphate is being replaced by ammonium nitrate and phosphate fertilisers contain less sulphur than previously, deficiencies may eventually appear here, but only if the degree of sulphur pollution in the atmosphere decreases markedly.

PHOSPHORUS

The quantity of phosphorus in the crop

UPTAKE AT HARVEST
Table 18 summarises recent experiments measuring the amount of phosphorus removed from the soil by sugar-beet crops. Warren and Johnson[360] found the uptake of phosphorus on Barnfield at Rothamsted was as little as 0·04 cwt/acre P_2O_5 in roots plus tops from plots which had received no fertiliser for over 100 years, and up to ten times as much from plots with fertiliser. Mattingly et al.[243] recorded similar results in an experiment started in 1899 at Saxmundham.

39

Widdowson et al.[371] measured uptake of phosphorus by sugar beet and other crops at Woburn for four years and found on average the sugar-beet roots contained 0·22 cwt/acre P_2O_5 and the tops 0·18 cwt/acre P_2O_5. The range over the four years was 0·12 to 0·34 cwt/acre in roots and 0·10 to 0·27 cwt/acre in tops. The total removed over this period by roots plus tops was 1·57 cwt/acre—greater than oats (1·06 cwt/acre), barley (0·53 cwt/acre) or potatoes (0·94 cwt/acre) but less than grass leys (1·84–2·14 cwt/acre).

TABLE 18

QUANTITY OF PHOSPHORUS IN THE SUGAR-BEET CROP AT HARVEST

	Range (cwt/acre P_2O_5)	Average crop with fertiliser	Range (kg/ha P)	Average crop with fertiliser	References
	0·05–0·41	0·35	3–22	19	Warren and Johnson[360]
	0·18–0·73	0·38	10–40	21	Goodman[132]
	0·14–0·71	0·45	8–39	25	James et al.[193]
	0·22–0·62	0·39	12–34	21	Widdowson et al.[371]
	0·11–0·47	0·43	6–26	24	Mattingly et al.[243]
	0·37–0·64	0·50	20–35	27	Draycott et al.[104]
Means	0·18–0·60	0·42	10–33	23	

In commercial practice, an average crop of sugar beet given a moderate dressing of phosphorus and other fertiliser elements contains 0·40–0·60 cwt/acre P_2O_5 but large crops given large dressings of phosphorus may contain up to 1·00 cwt/acre. Table 19 shows the amount of phosphorus in tops and roots separately. About half is in the tops and half in the roots so where tops are ploughed in, 0·2–0·3 cwt/acre is removed in the roots and 0·2–0·3 cwt/acre returned to the soil.

CONCENTRATION AT HARVEST

Table 19 also shows the concentration of phosphorus in the dry matter of tops and roots, which ranges at Broom's Barn from 0·18 % to 0·40 % in tops and from 0·09 % to 0·17 % in roots. An average crop given a moderate dressing of fertiliser contains 0·34 % P in tops and 0·15 % P in roots. Larsen[216] reported that sugar-beet petioles contained from 0·19 to 0·27 % P and laminae from 0·27 to 0·39 % P. In Russia, Gurevich and Boronina[140] found 0·15 % P in tops and 0·13 % P in roots. They showed that large amounts of fertiliser

TABLE 19

QUANTITY OF PHOSPHORUS AND CONCENTRATION IN THE CROP
AT HARVEST. BROOM'S BARN, 1965–70
(after Draycott et al.[104])

	Range	Average crop with fertiliser
	Quantity in the crop (cwt/acre P_2O_5)	
Tops	0·15–0·53	0·28
Roots	0·13–0·33	0·24
Total	0·28–0·86	0·52
	Quantity in the crop (kg/ha P)	
Tops	8–29	15
Roots	7–18	13
Total	15–47	28
	Concentration in dry matter (% P)	
Tops	0·18–0·40	0·34
Roots	0·09–0·17	0·15

phosphorus only slightly increased these values, particularly under
dry soil conditions.

CONCENTRATION AND UPTAKE DURING GROWTH
Knowles et al.[203] reported the concentration of phosphorus in sugar-
beet plants from emergence until harvest, and Table 20 shows the
concentration in both leaves and roots was large when the plant was

TABLE 20

CONCENTRATION OF PHOSPHORUS IN WHOLE SUGAR-BEET
PLANTS AND IN TOPS AND ROOTS FROM EMERGENCE TO HARVEST
(after Knowles et al.[203])

Date	Concentration in dry matter (% P)		
	Whole plant	Tops	Roots
31 May	0·65	0·73	0·21
21 June	0·51	0·53	0·41
5 July	0·45	0·44	0·48
19 July	0·32	0·35	0·28
1 August	0·32	0·34	0·29
16 August	0·32	0·34	0·30
31 August	0·27	0·33	0·24
13 September	0·15	0·21	0·13

young. The concentration decreased rapidly during June, more slowly in August and then rapidly until harvest. Gurevich and Boronina[140] also found the concentration decreased slowly at first and then rapidly until harvest in the Russian crop, from 0·32 in May,

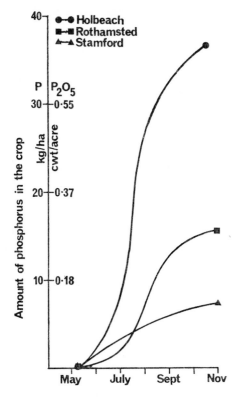

FIG. 10.   Total amount of phosphorus in sugar-beet crops on three contrasting soils.[132]

0·30 in June to 0·13% P in October. Goodman[132] analysed sugar-beet crops from three locations in England and found that the uptake of phosphorus varied greatly with soil (Fig. 10). On a shallow limestone soil the crop contained 0·18 cwt/acre P_2O_5, on a clay with flints (Rothamsted) 0·27 cwt/acre, and on a deep silty loam nearly 0·73 cwt/acre at the end of the season.

Phosphorus-deficiency symptoms

Phosphorus-deficiency symptoms on mature sugar beet are uncommon for they only appear when the concentration of available soil phosphorus is extremely small. The symptoms are more common on seedlings, especially where other factors such as soil acidity or root damage decrease uptake of the element. Irrespective of the age of the plant, phosphorus deficiency is typified by dark green leaves and stunting of the whole plant. Deficient plants look as if they were planted several weeks later than comparable plants with an adequate supply of phosphorus.

Seedlings on severely deficient soil die from attacks of black leg (*Pythium* spp.) and other fungal infections and it is difficult to establish a full plant stand. In N. America the main pathogen on phosphorus-deficient soils is *Aphanomyces cochlioides*. In the seedling stage, the cotyledons and primary leaves of phosphorus-deficient plants are dark green and then become pitted and may die. Older affected plants often show an abnormal amount of reddening, particularly when virus yellows is present. The petioles curl upwards and a blackish-brown necrosis develops from the tips and edges of the leaves without previous chlorosis.[187] In mature severely-deficient plants a brown netted veining forms in the tissues of the fully expanded leaves when they dry up and die, in contrast to the uniform yellowing which occurs in more deficient plants when leaves senesce.[347] Sipitanos and Ulrich[313] reported that deficient plants grown in nutrient culture developed golden areas on the leaves which became necrotic.

Root growth is also stunted by shortage of phosphorus and the tap root often forms a mass of fibrous secondary roots. Ulrich and Hills[347] draw attention to the effect of soil temperature on phosphorus absorption. The element is absorbed slowly when the temperature is low (50°F) and more rapidly as the temperature rises. At 68°F it is so rapid that leaf symptoms of deficiency usually disappear. During cold spring weather growth may be slowed by transient phosphorus deficiency without symptoms appearing on the leaves, and chemical analysis is needed to diagnose deficiency under these conditions.

Effect of phosphorus fertiliser on yield

In Great Britain

Mineral soils

Crowther and Yates[71] summarising responses to phosphorus by sugar beet up to 1940 found 1·0 cwt/acre P_2O_5 increased sugar yield on average by 1·6 cwt/acre in 200 experiments. Boyd et al.[32] found the response was only 1·1 cwt/acre in 200 experiments between 1934 and 1949. In the most recent group of experiments made in this country (78 experiments made between 1957 and 1960)[100] the average response to 1·0 cwt/acre P_2O_5 was of the same order, 1–1·5 cwt/acre sugar or about 3% of the mean yield. The average increase in yield was worth less than the cost of the fertiliser, but on some fields phosphorus fertiliser increased yield by over 10 cwt/acre sugar.

In ten experiments in Northern Ireland, McAllister and Rutherford[233] also found that 0·50 cwt/acre P_2O_5 increased sugar yield by only 1·9 cwt/acre and 1·00 cwt/acre increased it by a further 0·8 cwt/acre. However, Lee and Gallagher[222] found sugar beet gave remarkably large responses in Eire in 60 experiments, but the amount of phosphorus fertiliser needed for maximum yield was remarkably similar (1·4–1·6 cwt/acre P_2O_5) for all the different soil types studied.

Organic soils

In contrast to the small average responses described above, some soils show larger than average responses and sugar beet needs more fertiliser. Phosphorus given to sugar beet on organic fen soils in the last war increased sugar yield by 6·5 cwt/acre.[156] Post-war experiments confirmed this and 1·4–2·0 cwt/acre P_2O_5 was needed for maximum yield.[73] In 32 other experiments Boyd et al.[32] found the average response was 2·1 cwt/acre sugar.

Large dressings of fertiliser given to such fields have left considerable residues and the most recent experiments[101] indicate that responses are no greater than on mineral soils, for phosphorus only increased yield by 1·4 cwt/acre sugar. However, on deficient fields yield was increased by up to 17 cwt/acre. Adams[1] suggested giving an extra 0·50 cwt/acre P_2O_5 to these soils, also to chalky boulder clays and to newly ploughed grassland.

In Other Countries

Reports from many countries confirm these small responses to phosphorus fertiliser (Table 21) and the average percentage increase in yield is usually about 3–6% of the mean yield. Gericke[131] in Germany and James et al.[193] in Washington State, USA, chose deficient soils for experiments and the average increase in yield was much greater, being about 15% of the mean yield, but such responses are exceptional. On average, the amount of fertiliser needed for maximum yield in most countries is 0·40–0·80 cwt/acre P_2O_5 and only extremely deficient soils need more.

TABLE 21
EFFECT OF PHOSPHORUS FERTILISER ON ROOT YIELD

P dressing	Germany[131]	Great Britain[100]	Michigan USA[74]	Poland[209]	Mean
(cwt/acre P_2O_5)			(ton/acre)		
0	16·7	16·3	(14·7)	10·8	14·6
0·50	17·1	16·7	15·5	11·5	15·2
1·00	19·5	16·7	15·9	11·1	15·8
1·50	19·1	(16·7)	16·7	11·1	15·8
(kg/ha P)			(t/ha)		
0	42	41	(37)	27	37
28	43	42	39	29	38
56	49	42	40	28	40
84	48	(42)	42	28	40

Some values ( ) extrapolated and interpolated.

Effect of farmyard manure on response

Adams[8] tested response to 0, 0·5 and 1·0 cwt/acre P_2O_5 in the presence and absence of farmyard manure:

	P_2O_5 (cwt/acre)		
	0·0	0·5	1·0
	Sugar yield (cwt/acre)		
Without farmyard manure	51·8	52·9	53·1
With farmyard manure	54·3	54·8	55·3

He suggested that the best dressings for maximum yield were 0·5 cwt/acre P_2O_5 in the absence of manure, and none where manure was given. Draycott[84] confirmed these findings and showed that response to more than 0·3 cwt/acre P_2O_5 was negligible when 12 ton/acre farmyard manure was given (see page 133 for a more detailed account of the effects of farmyard manure on fertiliser requirement).

Residual phosphorus in soil

Residues of phosphorus from many years of fertiliser applications have been compared with fresh fertiliser given to plots which received little or no phosphorus. On the Exhaustion Land at Rothamsted, fresh phosphorus fertiliser gave yields which equalled those with residues.[66] On Agdell field at Rothamsted, fresh fertiliser did not, however, raise yields to those from plots with residues,[356] a result which confirmed the finding at Woburn[359] and of the Cockle Park experiment,[291] although sugar beet was not a test crop at Cockle Park. The difference between the two Rothamsted fields is thought to be because Agdell is a heavier soil, which is difficult to work.

Johnston et al.[195] reviewing the value of residues from superphosphate at Rothamsted and Woburn concluded that if a crop has to rely on newly applied phosphorus it must be accessible to the roots. Residues, which have been in the soil a long time, are intimately mixed with the soil, so the growing roots can get phosphorus from anywhere in the cultivated layer. Thus the plant is not prevented from obtaining enough by poor mixing of a fresh dressing, or poor soil structure restricting root growth to a limited volume of soil. Phosphorus residues are accumulating in most sugar-beet growing soils but how many soils would behave as those on the Exhaustion Land and those on Agdell and at Woburn is not known. More experiments are needed to set limits, on each type of soil, to which phosphorus residues must be accumulated before sugar beet gives no response to new phosphorus.

Efficiency of applied phosphorus

Olsen et al.[261] compared different forms of phosphorus fertiliser and found that the crop absorbed about 10 to 12% of the fertiliser phosphorus. When the phosphorus was placed near the seed, uptake by seedlings was greatly increased. Widdowson et al.[371] grew sugar beet with and without 0·50 cwt/acre P_2O_5 at Woburn and, on average of four years of cropping, the apparent recovery of fertiliser (weight in crop with fertiliser minus weight in crop without fertiliser) was up to 0·06 by tops and up to 0·03 cwt/acre P_2O_5 by roots. The average percentage recovery

$$\frac{\text{apparent recovery}}{\text{amount applied}} \times 100$$

was only 3% per year, which was small compared with recovery at Broom's Barn,[104] where the average recovery of phosphorus fertiliser by sugar beet given a commercial dressing is about 10%.

Effect of soil pH on response to phosphorus

With most crops (and sugar beet is no exception) responses to phosphorus fertiliser increase with increasing acidity of the soil. As sugar beet will not grow satisfactorily on acid soil, the effect is less noticeable than with crops which tolerate acid soil but even so, small changes in pH affect response:

pH	Increase in root yield from phosphorus fertiliser (ton/acre)
<6·5	1·03
6·5–7·0	0·68
>7·0	0·33

Variety and fertiliser requirement

Siwicki[314] found that different varieties of sugar beet in Poland needed different amounts of phosphorus fertiliser for maximum yield. The 'E' and the 'N' types made use of more fertiliser and gave a larger economic return from it than the 'Z' type (*see* page 26). In the same country, Lachowski[214] confirmed that the higher yielding the variety the larger the response to fertiliser and the more fertiliser required. These results need further investigation in other countries.

Effect of phosphorus on plant establishment

An adequate supply of soil phosphorus is needed to ensure that sufficient seedlings survive attacks by pests and diseases. In experiments testing phosphorus fertilisers, plots given none often yield less than fertilised plots because there are less roots at harvest, particularly on fields with small concentrations of soil phosphorus. Phosphorus fertiliser on deficient soils increases the vigour of sugar-beet seedlings,[75,290,313] which increases the number which survive.

Interaction with other fertilisers

Cope and Hunter[69] reviewed interactions between phosphorus and other fertilisers for many crops and concluded that for sugar beet, interactions were relatively unimportant. Many experiments in

Great Britain have measured interactions between phosphorus and other elements and most confirm this conclusion. Reviewing all the pre-war published information from Great Britain and Western Europe, Crowther and Yates[71] found a small positive interaction between nitrogen and phosphorus but none between potassium and phosphorus. Boyd et al.[32] examined response to 1·0 cwt/acre P_2O_5 in over 300 factorial experiments in Great Britain from 1934–49. On mineral soils it increased sugar yields by 1·6 cwt/acre sugar, the N × P interaction was +0·3 cwt/acre, the P × K interaction +0·2 and the N × P × K +0·1 cwt/acre.

Gallagher[124] in Eire found that the interaction between phosphorus and potassium was positive and statistically significant (confirmed by Fuehring et al.[123] in the Lebanon) although not as important as the N × K interaction. Jønsson[196] in Sweden investigated the N × P interaction and found that it did not affect yield significantly.

There is some evidence, where phosphorus is in very short supply, that interactions *are* important. Trist and Boyd[338] in Rotation I Experiment at Saxmundham showed that where soil phosphorus had been depleted, nitrogen fertiliser alone had little effect on yield but when phosphorus fertiliser was given, nitrogen increased yield greatly. Also Hills et al.[179] on phosphorus-deficient soil and in culture solution, showed that a shortage of phosphorus decreased the absorption of nitrate nitrogen. Gurevich and Boronina[140] in Russia found that nitrogen given to sugar beet only increased the percentage of nitrogen greatly when phosphorus was also given.

Suggestions that giving large dressings of phosphorus fertiliser might improve the quality of sugar beet given large dressings of nitrogen were ill-founded. Peterson et al.[275] found additional phosphorus fertiliser did not improve sugar percentage but neither did it do any harm. Ogden et al.[259] also found that large dressings of phosphorus did not offset the detrimental effects of excess nitrogen fertiliser.

Prediction of phosphorus requirement by plant analyses

Using similar techniques of plant sampling to those used for nitrogen determination, Ulrich and co-workers in USA have suggested phosphorus concentrations needed for maximum growth.[340,342,345] When dried petioles of recently matured leaves contained more than 750 ppm P soluble in 2% acetic acid the crop was not deficient in the element. On 17 fields examined, only one crop was deficient. It was established that 50 petioles from each field were needed to give a good estimate of the phosphorus status of the crop.

The value 750 ppm was shown by them and by Haddock and Stuart[147] to apply to crops from the singling stage until harvest. In order to determine the critical concentration for seedlings, Sipitanos and Ulrich[313] grew plants in water culture. The critical level in dry matter (where there was 10% decrease in yield of tops) for the cotyledons was 3 300 ppm soluble phosphorus (4 400 ppm total phosphorus) and 1 500 ppm for the petioles of older leaves (2 300 ppm total phosphorus).

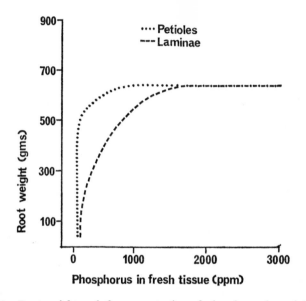

FIG. 11. Root weight and the concentration of phosphorus in petioles and laminae.[345]

Ulrich[345] examined the relationship between storage root growth and 2% acetic acid soluble phosphorus concentrations of petioles and blades of recently matured sugar-beet leaves (Fig. 11). Sugar beet on fields in California where the petiole phosphorus remained above 750 ppm for the entire season did not respond to phosphorus fertiliser even though more phosphorus was absorbed.

Relationship between plant and soil phosphorus

Haddock and Stuart[147] made a survey of sugar-beet crops in Western USA and found no fields were deficient in phosphorus on the basis of Ulrich's 'critical level' of 750 ppm soluble phosphorus in the sugar-beet petioles. These plant analyses were in good agreement with predicted results from sodium bicarbonate extractable soil phosphorus, for all the soils contained more than 15 ppm P. Only below 15 ppm P in the soil was a deficiency expected. James *et al.*[193] also showed that the concentration of phosphorus in

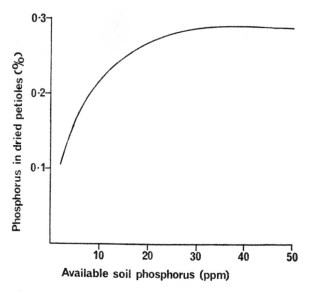

FIG. 12.   Sodium bicarbonate-soluble soil phosphorus and the concentration of phosphorus in dried petioles.[193]

petioles was related to sodium bicarbonate soluble soil phosphorus (Fig. 12). Only when the petiole phosphorus was less than 0·3% and the soil phosphorus less than 20 ppm did phosphorus fertiliser increase yield.

Prediction of phosphorus requirement by soil analysis

During the latter part of last century and for the first half of this century, 'available' soil phosphorus was commonly extracted by acidic solutions and acids of varying strength. Although some of these predicted response of certain crops with acceptable accuracy, none was satisfactory for sugar beet. Cooke[63] elucidated some of the problems involved. He showed that the available soil phosphorus was over-estimated if (1) iron and aluminium phosphates were dissolved, (2) too much extractant was used in relation to the weight of soil, (3) phosphate was dissolved from calcareous soils. Extracting soil phosphorus with 0·5 N acetic acid gave values which were poorly correlated with response by sugar beet to 6 cwt/acre superphosphate.

Other experiments using acid extractants with improved techniques were described later. Warren and Cooke[361] compared methods of analysis for soluble phosphorus in soils from 216 field experiments over eleven years. The experiments were divided by analysis into groups of equal numbers of fields, and crop responses used to decide the value of each analytical method. The best method

TABLE 22

MEAN RESPONSE TO PHOSPHORUS ON FIELDS GROUPED BY SOIL
ANALYSIS
(after Warren and Cooke[361])

P soluble in hydrochloric acid (ppm)	Number of fields	Sugar yield response to:	
		1·00 cwt/acre P_2O_5 (cwt/acre)	55 kg/ha P (t/ha)
2–15	27	5·0	0·63
16–25	26	2·7	0·34
27–36	27	1·8	0·23
38–47	27	1·2	0·15
48–59	27	0·9	0·11
61–80	27	0·6	0·08
82–114	27	1·0	0·13
115–350	28	0·4	0·05
Total/mean	216	1·7	0·21

was rapid extraction with dilute hydrochloric acid, but extracting with water or citric acid solution was nearly as effective. These three methods, using little solvent relative to soil, were more useful than methods using larger volumes of dilute sulphuric acid, dilute acetic acid or lactate solution. The mean response to 1·00 cwt/acre P_2O_5 on the 216 fields was 1·7 cwt/acre sugar, and the response exceeded 4 cwt/acre on only 35 fields. Table 22 shows the mean response for fields in eight groups with 27 fields in each group. Differential manuring of sites selected by soil analysis was more profitable than uniform manuring, with all the methods of soil analysis tested; the total amount of fertiliser used was the same with each method of manuring but the most efficient analytical method gave considerably more profit than uniform manuring.

Olsen et al.[262] proposed the use of sodium bicarbonate solution for extraction of plant-available phosphorus. This has been found to be more successful for the sugar-beet crop than acid extractants. For example, Draycott et al.[100] re-examined responses to phosphorus fertiliser in 78 experiments made between 1957 and 1960. Phosphorus was extracted from soil samples taken before the

TABLE 23

MEAN RESPONSE TO PHOSPHORUS FERTILISER IN RELATION TO SOIL PHOSPHORUS EXTRACTED BY SODIUM BICARBONATE AND OPTIMUM FERTILISER DRESSING

	Great Britain (after Draycott et al.[100]) ppm P in soil					
	≤ 10	11–15	16–25	26–45	≥ 46	
Increase in sugar yield from:						
1·00 cwt/acre P_2O_5	+8·6	+2·6	+1·5	+0·5	−1·0	(cwt/acre)
126 kg/ha P	+1·1	+0·3	+0·2	+0·1	−0·1	(t/ha)
Number of experiments	4	9	19	26	12	
Optimum fertiliser dressing:						
(cwt/acre P_2O_5)	1·50	1·00	0·50	0·25	0	
(kg/ha P)	180	120	60	30	0	

	USA (after Tolman et al.[336]) ppm P in soil				
	0–15	16–45	46–75	>75	
Increase in root yield	+2·4	+3·1	+2·5	+0·9	(cwt/acre)
	+0·3	+0·4	+0·3	+0·1	(t/ha)
Number of experiments	16	161	119	115	
Percentage of fields with response	100	74	72	33	
Optimum fertiliser dressing:					
(cwt/acre P_2O_5)		1·07	0·71	0·36	0
(kg/ha P)		60	40	20	0

experiments started, using four modern extractant solutions— sodium bicarbonate, anion resin, ammonium acetate/acetic acid and calcium chloride. The value of these four methods of analysis for predicting response to fertiliser were compared statistically. After allowing for variation in response due to experimental error, the percentages of the variance of the responses from field to field accounted for by each method were: sodium bicarbonate—40, anion resin—33, ammonium acetate/acetic acid—29 and calcium chloride—21.

Sodium bicarbonate solution is therefore recommended as the most useful extractant for sugar-beet soils which are necessarily neutral or alkaline (it was shown that prediction of response by ammonium acetate/acetic acid was poor on soils with much free calcium carbonate). The responses by sugar beet on fields with a range of soil phosphorus concentrations is shown in Table 23, together with the optimum fertiliser dressing.

Tolman et al.[336] made a similar study in several States of the USA; they compared sodium bicarbonate and carbon dioxide as extractants. The sodium bicarbonate method was generally better in predicting response. The lower part of Table 23 shows that sugar beet responded to phosphorus in the USA in a similar manner to response in Great Britain. Tolman et al.[336] recommend more phosphorus fertiliser for soils in the range 46–75 ppm P than Draycott et al.[100] In the range 0–45 ppm P there is very good agreement.

TABLE 24

EFFECT OF PHOSPHORUS FERTILISER ON SUGAR YIELD IN 29 EXPERIMENTS ON ORGANIC SOILS WITH SOILS GROUPED BY SODIUM BICARBONATE-SOLUBLE SOIL PHOSPHORUS AND BY LOSS ON IGNITION
(after Draycott and Durrant[101])

Soil P ($\mu g/ml$)	Number of fields	Response to fertiliser			
		(cwt/acre P_2O_5)		(kg/ha P)	
		0·75–0	1·50–0	41–0	82–0
		Sugar yield response			
		(cwt/acre)		(t/ha)	
7–20	6	+4·2	+5·2	+0·5	+0·7
21–35	7	+1·3	+0·9	+0·2	+0·1
36–45	7	+0·5	+0·7	+0·1	+0·1
>45	9	+0·3	−0·1	0	0

Loss on ignition (%)	Response to fertiliser	
	(cwt/acre P_2O_5) 1·50–0	(kg/ha P) 82–0
	Sugar yield response	
	(cwt/acre)	(t/ha)
>71	−0·8	−0·1
61–70	0	0
51–60	+1·3	+0·3
36–50	+1·0	+0·1
26–35	+4·9	+0·6
14–25	+1·2	+0·2

ORGANIC SOILS

Draycott and Durrant[101] investigated the phosphorus manuring of sugar beet on organic soils both in relation to soil phosphorus concentration (extracted with sodium bicarbonate) and to the amount of organic matter in the soil (measured by loss on ignition). Table 24 shows that the soil phosphorus concentrations were a useful guide to response. When the soil phosphorus was 7–20 µg/ml, 1·50 cwt/acre P_2O_5 was justified; when 21–45 µg/ml, only 0·75 cwt/acre was needed and when more than 45 µg/ml, no fertiliser was needed. Loss on ignition was less reliable in predicting response but responses were generally smaller the greater the loss on ignition.

Phosphorus residues and time of soil sampling

Mattingly et al.[243] used sodium bicarbonate to extract available phosphorus residues (from 70 years of manuring) from soils on Rotation II Experiment at Saxmundham. The plots were cropped with sugar beet and the yields were closely related to the soil phosphorus concentration (Table 25). Fresh superphosphate and residues

TABLE 25

SOIL PHOSPHORUS AND SUGAR-BEET YIELD AT SAXMUNDHAM, 1965–67

(after Mattingly et al.[243])

Manuring	Sodium bicarbonate- soluble P (ppm)	Roots (ton/acre)	Sugar (cwt/acre)	Roots (t/ha)	Sugar (t/ha)
None	4–6	8·3	26	10·4	3·3
FYM (1899–1964)	9–12	13·5	48	16·9	6·0
Ditto plus super- phosphate	20–40	16·9	59	21·2	7·4

together increased yield of roots by 1·7 ton/acre or 4 cwt/acre sugar—more than the residues alone.

The time of year when samples of soil are taken for analysis is important, for Garbouchev[125] showed that there is a seasonal fluctuation in soluble soil phosphorus under sugar beet. The concentration generally decreases during summer and autumn and increases again in winter. Differences between maximum and minimum concentration were large enough to be important in practice.

Changes in soil phosphorus in relation to applications and offtake

Sodium bicarbonate-soluble soil phosphorus values are useful both for predicting fertiliser requirement of crops on a year-to-year basis as already described, and for assessing changes over a period of years. Regular analyses allow fertiliser to be applied only when needed to maintain a satisfactory soil phosphorus concentration, *e.g.* once in a rotation to the most responsive crop. The amounts of fertiliser given during five years' cropping at Broom's Barn and the amounts of phosphorus removed in the crops in relation to the change in soil phosphorus were:

	P_2O_5 (cwt/acre)		Change in bicarbonate soluble P
Applied	Removed	Difference	(ppm)
0	1·0	−1·0	−5·5
2·0	1·3	+0·7	+0·5
4·0	1·4	+2·6	+13·5

Thus on a soil with a moderate supply of phosphorus (initially 35 ppm) large quantities must be removed or applied to change the soil concentration by a measurable amount. When the amount applied slightly exceeded that removed, soil phosphorus tended to decrease. Giving 2·6 cwt/acre more than was removed increased soil phosphorus by 8 ppm and giving 1·0 cwt/acre less than was removed decreased it by about 5 ppm.[104] This confirms the linear relationship proposed by Williams and Cooke[375] from soil analyses and assumed crop compositions in the Saxmundham Rotation I Experiment. Similar experiments are needed on different soils so that the amounts of fertiliser required to increase (or maintain) available soil phosphorus at predetermined concentrations satisfactory for maximum yield can be predicted.

SULPHUR

Deficiency symptoms

Ulrich and Hills[347] described symptoms of sulphur deficiency in sugar-beet crops in California, USA. Leaves first become yellow, old and young leaves alike (cf. nitrogen deficiency where the heart leaves remain green). As the deficiency becomes more severe, irregular brown blotches may appear on the leaf blades and on the petioles. The root system is affected little by a shortage of sulphur. Haddock and Stuart[147] found that sugar-beet plants in Western USA contained 3 to 7% sulphur in dried laminae. They reported that sulphur-deficiency symptoms were increasing in sugar-beet crops.

Sulphur in the crop

Whitehead[369] found that a 14 ton/acre crop of sugar beet removed 28 lb/acre sulphur. Because the ratio of nitrogen to sulphur in proteins is about 12:1, crops need about one-tenth to one-fifteenth as much sulphur as nitrogen. Ulrich and Hills[347] found that sugar-beet plants with sulphur-deficiency symptoms contained 50–200 ppm S (as SO_4) in dried leaf blades. Without symptoms, leaf blades contained from 500 to 14 000 ppm. The concentration below which the crop responded to treatment was 250 ppm.

Response to sulphur applications in the field

Atmospheric pollution over Great Britain supplies appreciable quantities of sulphur in rain (10–100 lb/acre/annum) and Cooke[65] suggested that the amount was large enough at present to satisfy crop requirements—certainly crops do not show signs of shortage. The decreased use of sulphur-containing fertilisers for sugar beet and other crops (e.g. ammonium sulphate and superphosphate) and decreased atmospheric pollution may eventually result in shortage of this element.

In California, USA, Ulrich et al.[344] described an experiment with sugar beet in which the crop showed symptoms of sulphur deficiency and responded to treatment with calcium sulphate. Without the sulphate the dried leaf blades contained only 750 ppm S (as SO_4) but with the treatment (484 lb/acre S as sulphate) it increased to 13 600 ppm.

Conclusions

Results of many experiments during the last ten years emphasise the small return from phosphorus fertiliser; arable soils where sugar beet has been grown for a long period usually contain sufficient residual phosphorus for the crop to yield fully without fresh fertiliser. On such soils future policy should be directed towards applying phosphorus fertiliser less frequently in the rotation, giving it to the most responsive crops and in sufficient quantity to prevent soil phosphorus from declining. Sugar beet removes up to 0·50 cwt/acre P_2O_5 from soil in tops plus roots, half of which is returned if the tops are ploughed in. A dressing of 0·50 cwt/acre P_2O_5 in fertiliser is justified where little is known about the reserves in the soil because it ensures no loss of yield through shortage of the element on nearly all fields and makes good the offtake.

Optimum phosphorus fertiliser dressings for sugar beet in relation to the amount of phosphorus extracted from a sample of soil by sodium bicarbonate solution and the ADAS soil index are as follows:

P in soil (ppm)	0–10	11–15	16–25	26–45	>45
Dressing (cwt/acre P_2O_5)	1·50	1·00	0·50	0·25	0
ADAS index	0	1	2	3	4

This method of analysis is a good guide to the amount of soil phosphorus available to sugar beet and other crops, and its adoption is recommended.

Where organic manure is given before sugar beet, phosphorus fertiliser can be omitted without loss of yield. As with other crops, soil pH affects availability of phosphorus to sugar beet and the smaller the pH the larger the response to phosphorus. The practical implication is to first ensure that pH is adequate for sugar beet (page 85ff). Soil texture affects response by some crops to phosphorus and more fertiliser is often given on the heavier-textured soils, but with sugar beet there is no evidence that texture affects response. Virgin organic soils are normally deficient in phosphorus because they lack minerals which release phosphorus during weathering. In Great Britain most of the fens have been adequately manured since the last war and sugar beet no longer gives very large responses to this element. The few fields which need large dressings can be distinguished by the soil analysis described above.

Chapter 4

Potassium and Sodium

Potassium and sodium are considered together because each greatly affects the requirements of the other by sugar beet. In Great Britain sodium is regarded as an important fertiliser element for sugar beet for, after nitrogen, an application of sodium increases sugar yield more than any other element. For example, on average of over 200 fields chosen at random, yield responses (cwt per acre) were: nitrogen +5·3, phosphorus +1·8, potassium +3·3, sodium +5·0. Not surprisingly, sodium fertiliser, usually as agricultural salt (crude sodium chloride) or kainit (mostly sodium and potassium chlorides (*see* page 103)), is considered necessary for maximum profit from sugar beet on most fields. Potassium has long been given, but it is only during the last thirty to forty years that the value of sodium has been recognised.

Experiments have shown that sodium and potassium fertilisers have similar effects on sugar-beet yield and that supplying one greatly decreases response to the other. The two elements are partly but not wholly interchangeable. In Great Britain sufficient rain usually falls in winter to leach most of the sodium released during weathering of minerals and that given as fertiliser, but in drier climates sodium is not removed and in some areas accumulates in toxic concentrations. In hot dry climates, particularly under irrigation with brackish water, sodium salts concentrate in surface soil as a consequence of upward water movement. As expected, sugar beet under such conditions gives no response to sodium fertiliser, but potassium is often needed.

POTASSIUM

Concentration of potassium in the crop

Table 26 shows the average concentration of potassium in dried sugar-beet tops at harvest is about 3 % whereas the concentration in roots is 0·77 %. The range of analyses shows that provided the crop is given potassium fertiliser the analysis of both tops and roots

TABLE 26
CONCENTRATION OF POTASSIUM IN SUGAR BEET AT HARVEST

| | Concentration in dry matter (% K) | | References |
	Tops	Roots	
	1·71–3·55	0·63–0·77	Johnston et al.[195]
	2·24	—	International Potash Institute[191]
	2·90–3·87	0·71–0·94	Draycott et al.[89]
	2·68–3·49	0·69–0·96	Draycott et al.[104]
Range	1·71–3·87	0·63–0·96	
Means	3·03	0·77	

is fairly consistent. Only when the crop is grown without fertiliser is the analysis of the tops decreased to less than 2% and the roots to less than 0·7%. When the potassium concentration in plant samples is determined throughout the growing season, the amount of potassium decreases rapidly during May and by July is at a similar concentration to that at harvest. Figure 13 shows the concentration of potassium in tops and roots and the quantity of the element in the crop throughout the growing period.

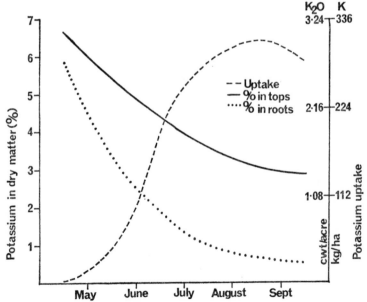

FIG. 13. Concentration of potassium in dry matter of tops and roots and the total amount in the crop.

TABLE 27

QUANTITY OF POTASSIUM IN THE SUGAR-BEET CROP AT HARVEST

	(cwt/acre K_2O)			Quantity in the crop (kg/ha K)			References
	Tops	Roots	Total	Tops	Roots	Total	
	0·38	2·76	0·95–2·91	40	288	99–303	International Potash Institute[191]
	0·79–2·05	0·36–0·83	1·17–2·88	82–214	38–87	122–300	Johnston et al[195]
	0·76–1·90	0·49–0·98	1·25–1·21	79–197	51–102	130–299	Draycott et al.[104]
	0·38–1·75	0·23–0·61	0·61–2·36	39–182	24–64	64–246	Widdowson et al.[371]
Range	0·38–2·05	0·23–2·76	0·61–3·14	39–214	24–288	64–328	
Means	1·14	0·89	1·67	119	93	174	

Quantity of potassium in the crop

Table 27 shows that when the crop is given potassium fertiliser the amount in tops and roots is often doubled. The amount in tops is usually considerably more than in the roots—approximately $1\frac{1}{2}$ times as much. This means that when the roots only are removed from the field and the tops ploughed in, the offtake of potassium is very much less than when both tops and roots are removed. Widdowson et al.[371] found that sugar beet recovered one-third of the potassium applied in fertiliser.

Table 28 shows the relationship between the maximum potassium and sodium uptake by sugar beet (usually in August or September)

TABLE 28

MAXIMUM QUANTITY OF POTASSIUM AND SODIUM IN SUGAR BEET (AUGUST/SEPTEMBER) AND QUANTITY IN THE CROP AT HARVEST
(after Durrant and Draycott[110])

	Tops only		Tops plus roots		Tops only		Tops plus roots	
	(cwt/acre)				(kg/ha)			
	K_2O	Na_2O	K_2O	Na_2O	K	Na	K	Na
Maximum	2·10	1·61	2·75	1·81	218	75	285	84
Quantity in crop at harvest	1·62	1·18	2·32	1·42	168	55	241	66
Proportion of maximum: quantity in the crop at harvest (%)					77	73	84	79

and the amount in the crop at harvest. In tops, the maximum uptake was over 2·10 cwt/acre K but only 1·62 cwt/acre at harvest. Similarly, when both tops and roots were considered the amount in the crop was nearly 2·75 cwt/acre in the summer but only 2·32 cwt/acre at harvest. Less sodium than potassium was taken up, but the ratios between maximum uptakes and uptakes at harvest were similar.

Figure 13 shows the uptake of potassium by the sugar-beet crop at Broom's Barn throughout the growing period. As the concentration in tops and roots declined the dry matter production increased rapidly until the autumn. In the last stages of growth before harvest, the leaves died and the total amount of potassium in the crop decreased slightly.

Potassium-deficiency symptoms

In the field

Hale *et al.*[150] described the field deficiency symptoms, often referred to as 'scorch'. The symptom which appears first is apical and marginal chlorosis with a dull olive green appearance of the leaves, usually in July but earlier in severe cases. The surface of the leaf becomes dull and bronze in colour and later small clusters of diffuse buff-coloured spots appear, generally arranged in the form of rough triangles, the bases of which are towards the margins of

TABLE 29

ANALYSIS OF SUGAR-BEET LEAVES WITH AND WITHOUT
POTASSIUM-DEFICIENCY SYMPTOMS

Concentration (% dry matter)	With symptoms	Without symptoms
	(after Hale *et al.*[150])	
	% in dried leaves	
K	0·61	1·69
Na	0·69	0·88
Ca	2·06	1·38
Mg	1·00	0·77
	(after Brown *et al.*[42])	
	% K in dried laminae	
Organic soil	0·6	2·3
Clay loam	1·8	2·6
	(after Ulrich and Hills[347])	
	% K in dried petioles	
>1·5% Na	0·2–0·6	1·0–11·0
<1·5% Na	0·5–2·0	2·5–9·0
	% K in dried laminae	
>1·5% Na	0·3–0·6	1·0–6·0
<1·5% Na	0·4–0·5	1·0–6·0

the leaf with the apices projecting inwards between the veins. On some plants there is a continuous band of shrunken chlorotic tissue round the margin of the leaf and the plants look stunted. Brown striped lesions commonly appear on the petioles. Table 29 shows the concentration of potassium in plants with and without symptoms of potassium deficiency.

The scorch symptoms develop after the chlorosis or may develop independently of it. Necrosis generally follows, both forming an unbroken border round the leaf and lobes between the veins. The necrotic tissue is dull or reddish brown, it is tough but soft to handle, and it does not crumble or disintegrate like the necrosis associated with magnesium deficiency. All the leaves on affected plants are thin and flaccid. The plant is normal in habit above ground but the tap root development is very poor. Plants affected early in the season are stunted but symptoms often do not occur until considerable growth has been made, and frequently affects bolters more severely than other plants. Hale *et al.*[150] considered that the deficiency symptoms were not always caused simply by shortage of potassium but also by shortage of sodium. They found that the symptoms were increased in severity by nitrogen (ammonium sulphate) and phosphorus (superphosphate) fertilisers and were cured in some cases by application of either sodium or potassium fertilisers, and in some cases by sodium only.

In sand cultures
Bernshtein and Okanenko[20] investigated the effect of potassium deficiency on photosynthesis, respiration and phosphorus metabolism in sand culture. Severe potassium deficiency increased the phosphorus concentration in leaves of young, but not in mature, plants. Severe potassium deficiency both increased the plant's rate of respiration and decreased its rate of photosynthesis. Ulrich and Hills[347] in recent sand culture experiments found that potassium was translocated from older to younger leaves and at times this was sufficient to provide for a considerable amount of new leaf growth. Sometimes the centre leaves developed calcium-deficiency symptoms and laminae with potassium-deficiency symptoms had a faint but distinct odour of decay. The upper side of leaves which were short of potassium had a dull surface owing to the disappearance of the waxy layer covering the epidermal cells, which may influence susceptibility to damage by pests and diseases.

The effect of potassium fertiliser on yield

Table 30 shows the average effect of potassium fertiliser on 42 fields of sugar beet[329] which are typical of many results from various parts of the world. Potassium increases root yield and sugar percentage, with consequent increased total sugar yield. Although potassium often increases yield of tops early in the growing period, at harvest the effect has disappeared. There is conflicting evidence of the effect of potassium on juice purity but in Tinker's experiments, as in many others in various parts of the world, potassium had no effect (*see* page 207 for the effect of potassium fertilisers on sugar-beet quality).

Boyd *et al.*[32] reported that the response to potassium in nearly 400 experiments from 1934 to 1949 was 2·13 cwt/acre of sugar. The effect of potassium was closely related to the amount of nitrogen applied and in the presence of nitrogen, dressings in excess of 1·2 cwt/acre K_2O gave a profitable increase in yield. Sugar beet on soil derived from the chalky boulder clay was exceptional in giving no response but apart from this, there were only small variations in responses to potassium between factory areas. An application of 5 cwt/acre NaCl gave a substantial response in most parts of the

TABLE 30

MEAN EFFECT OF POTASSIUM FERTILISER ON 42 FIELDS
(after Tinker[329])

	Without K	*With 1·2 cwt/acre K_2O* (125 kg/ha K)
Root yield (ton/acre)	16·7	+0·3
(t/ha)	41·9	+0·8
Tops yield (ton/acre)	12·2	−0·1
(t/ha)	30·6	−0·3
Sugar yield (cwt/acre)	54·8	+1·6
(t/ha)	6·88	+0·20
Sugar percentage	16·4	+0·2
Juice purity (%)	93·9	0

country, whether or not potassium was also applied, but responses to potassium were usually small when salt was applied. Response to potassium in 216 experiments between 1957 and 1969[111] was 4·8 cwt/acre of sugar in the absence of sodium but only 0·2–0·7 cwt/acre where 0·8 cwt/acre sodium was given.

INTERACTIONS

Nitrogen and phosphorus
Boyd *et al.*[32] described interactions between potassium and other elements in experiments made between 1934 and 1949. In the absence of sodium, the (potassium) × (nitrogen) interaction was +0·77 cwt/acre of sugar. The (phosphorus) × (potassium) interaction and the (nitrogen) × (phosphorus) × (potassium) interactions were +0·17 and +0·14 cwt/acre respectively, neither of which was significant. Boyd *et al.*[32] drew attention to the response to potassium alone, which was +1·27 cwt/acre sugar compared with response to potassium in the presence of other nutrients, which was +3·15 cwt/acre.

TABLE 31

EFFECT OF POTASSIUM AND NITROGEN FERTILISER ON SUGAR
YIELD: MEAN OF 42 FIELDS
(after Tinker[329])

K dressing (cwt/acre K₂O)	(kg/ha K)	N dressing			
		(cwt/acre)		(kg/ha)	
		0·6	1·2	75	150
		Sugar yield			
		(cwt/acre)		(t/ha)	
0	0	54·5	55·1	6·84	6·92
1·2	125	55·8	57·0	7·00	7·15
2·4	250	56·3	57·7	7·07	7·24

Tinker[329] in later experiments investigated the (nitrogen) × (potassium) interaction and Table 31 shows the sugar yields with two amounts of nitrogen and three amounts of potassium. The positive interaction between these two elements was significant.

Magnesium
Draycott and Durrant[83] reported that potassium fertiliser increased the severity of the symptoms of magnesium deficiency on sugar-beet leaves grown on magnesium-deficient soils. Sugar yields shown in Table 32 from plots with two amounts of potassium in the presence and absence of magnesium fertiliser show that both elements were needed for maximum yield and that they did not interact. Consequently, they recommend that even where magnesium-deficiency symptoms are present the sugar beet should receive potassium together with magnesium fertiliser.

TABLE 32

EFFECT OF POTASSIUM AND MAGNESIUM FERTILISER ON SUGAR
YIELD: MEAN OF 19 FIELDS
(after Draycott and Durrant[83])

Mg dressing (cwt/acre)	(kg/ha)	K dressing			
		(cwt/acre K₂O)		(kg/ha K)	
		1·00	2·00	104	208
		Sugar yield			
		(cwt/acre)		(t/ha)	
0	0	49·7	51·7	6·24	6·49
0·80	100	53·2	54·5	6·68	6·84

THE SIGNIFICANCE OF THE POTASSIUM:NITROGEN RATIO IN SUGAR-BEET FERTILISERS

Several investigators have examined the ratio of nitrogen to potassium in fertiliser which is best suited for maximum sugar yield and quality of the crop. Von Müller et al.[352] made experiments in sand culture to determine the effect of changes in the N-to-K_2O ratio on the development and yield of sugar beet. With a large supply of nitrogen the N-to-K_2O ratio was important and for maximum sugar yield a ratio of at least 1:3 was thought desirable. Under field conditions, where the soil is rich in nitrogen and short of potassium, the authors suggest that the ratio of N-to-K_2O in the fertiliser is critical, but no field experiments were reported.

Heistermann[166] made field experiments to determine the influence of changes in the N-to-K_2O ratio on the yield and sugar percentage of roots. The maintenance of a large dressing of nitrogen fertiliser was indispensable for good growth and this made the potassium fertiliser all the more important. With a wide N-to-K_2O ratio, e.g. 1:2 or 1:3, large dry matter yields were obtained, resulting in large yields of roots and sugar percentage. With the N-to-K_2O ratio 1:3 it was possible to decrease the glutamine content (the main component of the harmful nitrogen in the sugar-beet juice), with simultaneous large nitrogen fertiliser doses. The concentration of potassium was increased by the potassium fertiliser but the concentration of other elements was decreased.

Effect of soil type on response to potassium

Soil type greatly affects both the size of the increase in yield and the amount of potassium fertiliser needed for maximum yield. Clay soils, particularly the chalky boulder clays (i.e. calcareous soils of glacial origin), give least response, for weathering minerals and clay release much potassium. Organic soils also contain much potassium from decaying organic matter but sandy soils have little reserve of 'natural' potassium, and sugar beet gives most return and needs most fertiliser on these soils.[32,111] The mean increases in sugar yield (cwt/acre) from potassium in the absence of sodium fertiliser in 140 experiments were:

Sands   +9·2
Loams and clays +4·4
Organic soils   +0·4

Lee and Gallagher[222] examined the response to potassium by sugar beet grown on several soil types in Ireland. The percentage response to potassium was comparable on Carboniferous Limestone and Old Red Sandstone soils but the latter on average required 0·77 cwt/acre K_2O less than the former for maximum yield. The smaller potassium requirement of the Old Red Sandstone was probably due to the large amount of illite clay present, which releases much potassium. A relatively large percentage response to potassium on some of the Carboniferous Limestone soils compared with response on similar soils formed on gravel was attributed to the latter's poor capacity to release soil potassium. The presence of much free calcium carbonate was also thought to depress uptake of potassium.

Available soil potassium and response to fertiliser

Warren and Cooke[361] described 11 years of field experiments by E. M. Crowther to compare methods of analysing soils for available potassium. The experiments were divided by soil analysis into groups of equal numbers of fields and average crop responses were used to value the analytical methods. Such tables of average data over-value soil analysis because each method was misleading in a small proportion of the fields used (*see* Table 33). A quantitative way was developed of assessing the gains from using soil analysis in planning fertilising and of comparing analytical methods. The total profit from uniformly manuring all the soils was compared with the profit made by using the analysis to select a proportion only of the soils to be manured. The total amount of fertiliser used was the same with each manuring plan; the most efficient analytical method gave the most profit.

Citric acid used to extract available potassium separated the soils into groups for differential manuring, which was more profitable than giving uniform dressings to all fields. Acetic acid was less effective than citric acid; using hydrochloric acid to group the soils for differential manuring gave no advantage over uniform manuring with a large dressing of potassium. Water-soluble potassium measurements were of even less value than acid-soluble values. Using them to predict responses and manuring would have given less profit than uniform manuring with a large dressing.

Davis *et al.*[74] also used dilute HCl to analyse soils for available potassium in Michigan State, USA. The potassium was extracted by 0·135 N HCl with a soil-to-solution ratio of 1:4. The analyses were made on soil samples taken in August from plots of sugar beet where between none and 3·6 cwt/acre K_2O had been applied. There was a close relationship between the amount extractable in the

TABLE 33

RESPONSE TO POTASSIUM FERTILISER BY SUGAR BEET ON FIELDS GROUPED BY SOIL ANALYSIS
(after Warren and Cooke[361])

Soil K soluble in citric acid (ppm)	No. of fields	Mean response sugar (cwt/acre)	(t/ha)	Large 18·5–5·5 2·32–0·69	Medium 5·4–2·4 0·68–0·30	Small Sugar yield 2·3–0·1 0·29–0·13	Depressions 0·0– −8·2 (cwt/acre) 0·0– −1·03 (t/ha)
					Number of fields in each group		
26–41	30	4·6	0·57	11	12	5	2
42–51	33	4·8	0·60	16	9	3	5
52–57	31	6·2	0·78	16	10	4	1
58–66	30	2·7	0·34	6	10	6	8
67–75	31	2·1	0·26	5	8	11	7
76–95	31	2·0	0·25	3	11	11	6
96–127	32	0·8	0·10	2	3	15	12
128–218	30	0·7	0·09	4	2	10	14

hydrochloric acid and the amount applied to the soil. The sugar-beet yields showed that about 3·6 cwt/acre K_2O was needed for maximum yield.

Warren and Johnston[360] analysed soils from Barnfield experiment and found that the amount of potassium soluble in ammonium acetate was related to the amount of potassium in the roots of mangolds. They found that with large rates of fertiliser the potassium applied exceeded the amount taken up by crops. Not all the extra potassium from repeated dressings accumulated in the surface soil, for part moved down into the subsoil and some was lost in drainage. They suggest a 'saturation value' for soil. When more than this amount is applied then leaching is rapid. On Barnfield the saturation value was of the order of 300 ppm readily soluble K. Most potassium accumulated where no nitrogen fertiliser was given.

On the Barnfield plots the average increase to a depth of 21 in was 450 ppm exchangeable K where both farmyard manure and fertiliser had been given. This was only one-seventh of the potassium applied in 100 years of manuring and the difference was the loss by fixation and the loss in drainage water. The Barnfield soils had an exceptionally wide range of exchangeable potassium concentrations, approximately 90 to 900 ppm K. In soil:water suspensions, the proportion of exchangeable K released to water was found to be nearly constant for the full range of exchangeable K values, with the exception of soils with exchangeable K values less than 170 ppm, where the ratio of water-soluble to ammonium acetate–soluble potassium was smaller than for other soils.

Adams[7] made some interesting observations about the rôle of soil potassium analysis in predicting the amount of fertiliser needed by sugar beet. He pointed out that compound fertilisers which contained little potassium were used on sugar beet during the last war but since then there has been a change towards compounds containing larger amounts of potassium. Farmers resisted the idea that compound fertiliser with little potassium was suitable for sugar beet, thinking that most soils in eastern England were inherently deficient in potassium. This idea arose from the results of soil analysis. Exchangeable potassium measurements on soils used for fertiliser experiments during the late 1950's where potassium was measured in normal ammonium nitrate leachate, using the standards of the National Agricultural Advisory Service, showed that approximately 80% of the soils were either 'low' or 'very low' in potassium. The soils were a fairly representative sample of the sugar-beet growing areas. In spite of so many soils with small concentrations of potassium, Adams pointed out that sugar beet rarely showed obvious signs of potassium deficiency and that responses to potassium fertiliser were rare when sodium was given, there being little evidence of residual effects of potassium on the succeeding crops.

Adams pointed out that when Boyd et al.[32] and Warren and Cooke[361] examined the relationship between soil analysis and response to potassium they considered only the responses to potassium where sodium was not given. As there was no evidence that soil analysis could predict the soils on which potassium gave an economic response when sodium was applied, Adams considered that soil analysis for potassium was of little value in deciding the needs of sugar beet. The designation of many soils as 'low' or 'very low' in potassium simply encouraged the wasteful use of fertiliser, for much sugar beet received 10 cwt/acre of a compound fertiliser containing 18% K_2O, plus 6 cwt/acre of kainit containing 17–20% K_2O. In a later paper, Adams[4] tested sodium and potassium fertilisers on sugar beet on Lincolnshire limestone soils. The response to fertiliser varied widely in the different experiments and exchangeable soil potassium extracted by N/10 nitric acid, although not by N ammonium nitrate, predicted potassium response moderately well.

Tinker[334] leached organic soils with acetic acid and determined the potassium concentration in the leachate. The soils were from fertiliser experiments with sugar beet made in 1963–65. The results were compared with similar soil analyses for potassium described by Boyd et al.[32] on similar experiments made in 1934–1949. The potassium concentration in Tinker's soils was 420 ppm ± 37, and in Boyd's experiments 177 ppm ± 26. The very large increase in the mean value during the period probably explains the small response to potassium in Tinker's experiments. He suggested that this increase was due to large fertiliser dressings during the period which had accumulated as residues in the soil.

SODIUM

Amount of sodium in the crop

AT HARVEST

Table 34 shows the concentration of sodium in tops and roots of sugar beet at harvest and the quantity in the crop. The concentration in the tops is up to twenty times greater than in the roots. If the roots only are removed from the field, nine-tenths of the total plant sodium is returned to the soil. A mature crop of sugar beet contains about 0·45 cwt/acre Na, 0·05 cwt/acre in the roots and 0·40 cwt/acre in the tops.

TABLE 34

CONCENTRATION AND QUANTITY OF SODIUM IN SUGAR BEET AT HARVEST

| Na concentration (% dry matter) | | Quantity Na in the crop | | | | | | References |
| | | (cwt/acre) | | | (kg/ha) | | | |
Tops	Roots	Tops	Roots	Total	Tops	Roots	Total	
1·6-1·7	0·07-0·11	0·38-0·75	0·04-0·10	0·42-0·85	47-94	5-12	52-106	Draycott et al.[104]
0·4-1·7	0·08-0·10	0·05-0·41	0·02-0·10	0·07-0·52	6-51	3-13	9-65	Harmer et al.[158]
				0·17-0·66			21-83	Warren and Johnston[360]
0·96-1·04	0·04-0·06			0·40-0·48			50-60	Draycott and Farley[102]
0·74-1·48	0·04-0·11			0·48			60	Adams[5]
Range								
0·4-1·7	0·04-0·11	0·05-0·75	0·02-0·10	0·07-0·85	6-94	3-13	9-106	
Means								
1·2	0·08	0·40	0·06	0·45	50	8	57	

EFFECT OF SODIUM ON THE GROWTH AND NUTRIENT UPTAKE BY THE CROP

Adams[5] reported analyses made by J. B. Hale in 1941/42 at Rothamsted to compare the effects of sodium and potassium fertilisers on nutrient uptake. Sodium increased root yield in both years, but did not act by mobilising soil potassium reserves and increasing the potassium status of the plant, as had been suggested. Potassium fertiliser, although increasing the concentration of potassium in the plant, did not increase yield. Sodium and potassium were distributed differently in the plant and at harvest, only 6% of the total sodium was in the root compared with 33% of the potassium. The conclusion was that sodium was a nutrient for sugar beet and not a potassium substitute.

Tinker[89] made similar experiments at Broom's Barn, where the soil is much lighter than that at Rothamsted, testing a wider range of sodium and potassium dressings. Samples of the crop taken in summer and at harvest confirmed Adams' finding that it contained most sodium in August, much of which was in the tops, and that the total amount had decreased considerably by harvest. Periodic soil samples from the experiments showed that the amount of sodium in the crop was balanced by a corresponding decrease in the exchangeable soil sodium and in August the crop contained about 1·25 cwt/acre Na. Corresponding calculations for potassium in the crop not given it, and the decrease in exchangeable potassium in the soil, were not as closely related; plant uptake of 1·42 cwt/acre decreased the exchangeable potassium in the top soil by only 0·45 cwt/acre. When the crop was given fertiliser, the plant uptake and soil depletion were in good agreement—1·96 and 1·80 cwt/acre respectively. The stability of the exchangeable soil potassium in plots not given fertiliser may have reflected the transfer of potassium to and from a non-exchangeable pool or uptake from the subsoil.

More recently, Draycott and Farley[102] analysed the growth and nutrient uptake of sugar beet, with and without sodium fertiliser, from the early seedling stage until late harvest when sugar accumulation had ceased. Sodium increased the dry matter yield of tops and roots throughout the whole growing period. It also increased the sugar yield at each of three harvests in October to December but the size of the increase was about the same on each occasion. Sodium appeared to increase the sugar yield by several independent effects. Early in the year it greatly increased the leaf area index, which coincided with maximum solar radiation and day length (the number of leaves per plant was unaffected but the area of each leaf was increased). Another mechanism by which sodium increased sugar yield was by increasing the proportion of the total dry matter which was partitioned to the roots. In increasing the amount of dry matter in the roots it increased the yield of sugar, for root dry matter yield and sugar yield are very closely and positively correlated.[103] Sodium also improved the sugar percentage of fresh roots.

An irrigation treatment in the same experiments appeared to discredit the theory that sodium acts by improving the plant/soil water relationships. Plants given sodium are thought to wilt less readily than ones without sodium fertiliser. In two very dry summers at Broom's Barn when plants without irrigation wilted severely, observations did not substantiate this theory. The growth analysis and the sugar yields showed that sodium and irrigation had independent effects on the crop and there was no evidence for a negative interaction between them.

Sodium-deficiency symptoms

Although the leaves of healthy sugar-beet plants contain large quantities of sodium, plants grown without the element do not show any characteristic deficiency symptoms. The presence or absence of sodium in the nutrient medium does, however, influence the degree to which sugar-beet leaves show symptoms of potassium deficiency. When plants are deficient in both elements, potassium-deficiency symptoms are very severe. When plants are deficient in potassium but adequately supplied with sodium, potassium-deficiency symptoms are decreased in intensity and the growth of tops and roots is improved; instead of the severe interveinal scorch associated with potassium deficiency, symptoms are usually confined to marginal browning.[150,347]

Experiments comparing sodium and potassium fertilisers

IN GREAT BRITAIN
The first report of a comprehensive group of field experiments was that of Crowther.[72] In 152 experiments on commercial farms throughout the sugar-beet growing regions, four treatments were tested—no potassium, no sodium, potassium as 1·2 cwt/acre K_2O as potassium chloride, sodium as 5 cwt/acre agricultural salt (NaCl), and both sodium and potassium. On average of all experiments on mineral soils, the potassium gave 2·8 cwt/acre of extra sugar, an increase of 7%: 5 cwt/acre NaCl gave 5·1 cwt/acre of extra sugar, an increase of 16%. Taking into account relative prices for fertilisers and sugar beet at that time, the potassium gave an extra profit of 180% of the outlay, and the sodium an average profit of 540%. Crowther made the statement which has often been repeated,

"agricultural salt used at the rate of a few cwt/acre gives its own weight of extra sugar". The effects of sodium were very consistent from year to year, far more so than those of potassium. The sodium gave best results on sandy soils, especially those with small amounts of exchangeable potassium, but on organic soils sodium gave very little increase in yield. Where potassium was given there was a large additional response (2·7 cwt/acre of extra sugar) from using sodium as well, but where salt was given there was little benefit (0·4 cwt/acre extra sugar) from using potassium.

Boyd et al.[32] also reported on these and other experiments on over 200 fields. Besides testing nitrogen, phosphorus and potassium, they tested 0 and 5 cwt/acre NaCl in 190 experiments on mineral soils and 18 on organic soils. The mean responses are shown in Table 35. The value of sodium was clearly demonstrated, also the

TABLE 35

MEAN RESPONSES TO POTASSIUM AND SODIUM ON MINERAL
SOILS: 190 EXPERIMENTS
(after Boyd et al.[32])

	Sugar yield response to:					
	(cwt/acre)			(t/ha)		
	1·2 cwt/acre K₂O	5 cwt/acre NaCl	1·2 cwt/acre K₂O +5 cwt/acre NaCl	125 kg/ha K	250 kg/ha Na	125 kg/ha +250 kg/ha Na
No N or P	+1·2	+2·6	+2·7	+0·15	+0·33	+0·34
N only	+2·6	+4·8	+5·1	+0·33	+0·60	+0·64
P only	+1·6	+2·5	+2·3	+0·20	+0·31	+0·29
N and P together	+3·5	+5·0	+5·9	+0·44	+0·63	+0·74

large negative interaction between sodium and potassium. The response to potassium was scarcely sufficient to do more than repay its cost when sodium was given, but sodium gave an economic response whether or not potassium was applied.

EUROPE AND ASIA

Experiments at Göttingen, Germany[256] and on other experimental and commercial farms showed the greatest yields were always from crops given sodium and a small application of potassium. Large applications of potassium decreased yield slightly when sufficient sodium was given. In Holland, Lehr and Bussink[223] made experiments on soils with exchangeable sodium concentrations of 5–20 ppm Na. Sodium nitrate increased yields of sugar compared with calcium ammonium nitrate on all the fields and a considerable saving in potassium fertiliser was possible. The optimum dressings of potassium fertiliser were:

	With sodium nitrate	With calcium ammonium nitrate
	(cwt/acre K_2O)	
With FYM	0·96	1·93
Without FYM	1·93	2·90

Hernando et al.[173] compared four times of application of Chilean nitrate as a top dressing for sugar beet grown in Spain. They found that 1·20 cwt/acre given at thinning, ridging and in mid-July gave a larger yield than a single application, but giving the nitrate in August increased yield little. In Japan, Yasuda et al.[381] found that even with a sufficient supply of potassium, sodium increased yield by 3·7%. The amount of sodium needed was 0·48 cwt/acre Na as fertiliser, equivalent to 1·8 cwt/acre Chilean nitrate.

USA

In New Jersey, Leonard and Bear[224] showed that sodium was a useful fertiliser for sugar beet, red beet and other crops. They found that 0·90 cwt/acre Na applied as NaCl greatly increased yield both with, and particularly without, potassium fertiliser. The sodium concentration in the New Jersey soils was 20–40 ppm Na. A leaching experiment showed that sodium fertilizer equivalent to 20 years applications did not increase the exchangeable sodium in the soil sufficiently to damage its physical condition.

Harmer et al.[158] studied the value of sodium as sodium chloride and as a mixed sodium/potassium salt containing 56% NaCl for many crops grown on organic soils in Michigan. The sodium concentration in the soils was variable but sugar-beet yield was consistently increased by both forms of fertiliser:

	Yield of roots, 6-year average (ton/acre)
No sodium	14·5
4·5 cwt/acre NaCl	15·8
7·6 cwt/acre NaCl/KCl	16·0

Truog *et al.*[339] investigated the substitution of sodium for potassium in fertiliser applications for sugar beet on five Wisconsin soils which contained from 0·45–1·70 cwt/acre and 0·22–0·40 cwt/acre exchangeable potassium and sodium respectively. On four soils, sodium increased yield, particularly with small dressings of potassium but the yield was not increased by sodium on the soil containing the largest amount of sodium. Even with adequate or large amounts of available potassium the sodium produced a more vigorous crop and appreciable increases in yield and sugar percentage.

Considerably larger soil sodium concentrations were reported from California[147] for soils collected from heavy yielding sugar-beet fields contained 70 to 1 200 ppm exchangeable sodium. This is equivalent to 1·87 to 32·2 cwt/acre Na in the plough-layer of 9 in depth, so response to sodium fertiliser is extremely unlikely (*see* page 78 on soil analysis). Some sugar-beet fields there are under-drained so that they can be flooded from time to time to leach out the sodium salts that accumulate.

Growth of sugar beet in saline soils

In the sugar-beet growing regions where the annual precipitation exceeds evaporation and transpiration, sodium salts are rapidly leached from soil so residues of sodium fertiliser, sodium released from minerals during weathering and from organic matter do not accumulate. In climates where evaporation exceeds precipitation sodium does accumulate and the concentration of sodium is often large enough to damage crops. Irrigation with brackish water accentuates the problem.

The salinity of soils is a problem in many tropical and subtropical regions of the world; Kanwar[201] reported that 15 million acres of cultivated land in India alone are affected, and suggested that sugar beet would be a suitable crop to grow under these conditions, having shown that it was the most tolerant of several crops tested. Heald *et al.*[163] investigated the problems of saline soils in Washington State, USA, and found that sugar beet was one of the few crops which could be grown satisfactorily in saline conditions but that it was difficult to obtain a uniform plant stand. The most critical period in the growth of plants in saline soil was at germination and during emergence. The salt concentration in the Washington soils was 2 000 to 10 000 ppm or, expressed as electrical conductivity, 3 to 14 m mhos/cm. Pre-emergence irrigation increased plant stands from 20–30 plants/100 ft of row to 70–100 plants. The yield of roots was doubled by up to 24 in of flood irrigation before planting, which leached some of the sodium from the surface soil.

Use of sodium salts to control weeds in sugar beet

Coombe and Dundas[68] reviewed the herbicidal effect of sodium nitrate and chloride sprays in sugar-beet fields. They found that post-emergence spraying with 2½–3 cwt/acre sodium nitrate in 65–100 gallons of water was an effective herbicide, the best time of application being when the crop had two true leaves and the weeds were not beyond the cotyledon or small rosette stage. If preceded by good growing weather and followed by at least 24 hours of dry warm weather the spray was very effective. Some weeds were killed and the growth of others was checked; the authors give comprehensive details of the spectrum of weeds killed and the degree of resistance of many species. In addition to its action as a herbicide, the spray supplies nutrients for the sugar beet and the amount of fertiliser given before sowing can be decreased accordingly.

Selection for small sodium concentration

Several investigations[79,44] have shown that the sodium concentration of sugar-beet breeding lines is negatively correlated with sugar percentage. However, Finkner and Bauserman[120] selected pairs of roots with the same sugar percentage but with one having a large sodium concentration and the other having a small sodium concentration. The progenies bred true in that they both had the same sugar percentage, but the sodium percentages were quite different. This indicates that the sodium concentration itself had little effect on the sugar percentage and that it is not a useful criterion of sugar percentage in a breeding programme. However, as sodium is an important impurity in sugar-beet roots during processing, attention is being given to breeding for small sodium concentration.

Amount of sodium fertiliser needed by sugar beet for maximum yield

Few experiments have been made to determine the precise amount of sodium fertiliser needed by sugar beet for maximum yield. In Great Britain as in most countries, sodium in agricultural salt (NaCl) is inexpensive compared with most other plant nutrients. The policy in most experiments has therefore been to test only one fairly large dressing (often 5 cwt/acre NaCl) on the assumption that where sodium is applied for sugar beet, a sufficiently large dressing will be used to ensure maximum response.

Crowther[72] compared 3 and 6 cwt/acre agricultural salt. The larger dressing gave slightly more yield (0·7 cwt/acre sugar) than the small dressing but the increase was not statistically significant. Tinker[329] tested 0, 0·8 and 1·6 cwt/acre Na (2 and 4 cwt/acre NaCl) in 42 experiments. The sugar yields on average were 54, 57 and 58 cwt/acre respectively and it was concluded that about 1·2 cwt/acre Na or 3 cwt/acre NaCl was sufficient.

Soil analysis and prediction of response to sodium

There is little information about the relationship of field responses to sodium and the amount of sodium in the soil or in the plant. Adams[4] found no relation between field response to sodium and the amount of sodium extracted from soil with ammonium nitrate or N/10 nitric acid, but an inverse relation between response to sodium and exchangeable soil potassium. Tinker[331] measured response to sodium in 13 field experiments in 1965, the uptake of sodium by sugar beet in pots of the same 13 soils, and the exchangeable sodium in samples of the soils taken before sowing the crop. As sugar beet only responded significantly to sodium on two of the 13 fields, no clear relationship between response and soil analysis was found. However, it was established that the exchangeable sodium was very weakly held by the soil colloids. Nearly all of it was taken up by the plants in the pot experiments and uptake was closely related to the soil analysis.

Draycott[84] investigated the value of soil exchangeable sodium analyses for predicting requirement of sodium fertiliser on fields given a dressing of potassium (1·00 cwt/acre K_2O). All the significant responses to sodium were on soils with less than 25 ppm Na but sugar beet on several fields, particularly silty soils, gave no response even when the soil sodium value was small.

Holmes et al.[182] in Scotland and Adams[4] in England drew attention to the close correlation between response to sodium and exchangeable soil potassium. Where soil potassium was small, response to sodium was large, particularly where no potassium was given. Table 36 shows the results of the Scottish experiments. The soil potassium concentrations were designated small, medium and large, and responses to sodium were marked where the soil contained little exchangeable potassium. With a good supply of soil potassium, responses to both elements were small. Exchangeable soil sodium analyses (ammonium acetate/acetic acid) made on samples taken at the end of the experiments (up to eight years had elapsed) were not related to response to sodium.

TABLE 36

RESPONSE TO SODIUM AND POTASSIUM SEPARATELY AND
TOGETHER ON SCOTTISH SOILS GROUPED BY SOIL POTASSIUM
ANALYSIS

(after Holmes *et al.*[182])

Exchangeable potassium	No. of fields	No Na or K	Sugar yield		
			K without Na	Na without K	Na + K
			(cwt/acre)		
Small	3	35·4	+3·2	+6·7	+6·7
Medium	4	33·2	+3·2	+5·4	+7·2
Large	8	36·6	+1·7	+1·8	+1·3
			(t/ha)		
Small	3	4·44	+0·40	+0·84	+0·84
Medium	4	4·17	+0·40	+0·68	+0·90
Large	8	4·59	+0·21	+0·23	+0·16

Response to sodium on different soils in the United Kingdom

Boyd *et al.*[32] reporting on 200 experiments in the 1930's and 1940's
found little difference in response to sodium between soil types except
that sugar beet in 18 experiments on organic soils responded less
than on mineral soils. Durrant *et al.*[111] have investigated the effect
of soil type on response to sodium in experiments during the last
15 years. Sodium increased sugar-beet yield most on sandy-textured
soils both with and without potassium, whereas the crops on clay
soils gave little return from either element. Sodium had little effect on
yield in organic soils and was also of doubtful value on the silts
round the Wash and the Humber.

Loss of sodium from soil

That sodium does not accumulate in soils in Great Britain is well-
illustrated by analyses on Barnfield,[360] for applying sodium annually
for a century only increased the amount in the soil by 10 ppm Na
exchangeable in ammonium acetate. Williams[374] measured sodium
concentration in drainage water at Saxmundham and Woburn; at
Saxmundham it was 7–45 ppm Na and 20 ppm on average, and at
Woburn it was 7–28 ppm and 11 ppm on average. Tinker[332]
reviewed the factors affecting the movement of sodium in soils in
the United Kingdom, and from theoretical considerations and experimental results found that sodium applied in fertiliser was rapidly

leached from the soil even in the drier areas of the country. Winter rainfall was sufficient to remove most sodium applied in fertilisers in the course of two years.

Walsh,[355] reviewing sodium in Irish soil in relation to deposition in rain and response by sugar beet, found soils contained 70–350 ppm Na because annual deposition was from 0·36–2·68 cwt/acre (at Broom's Barn rainfall deposits less than 0·09 cwt/acre/annum Na).[89] Walsh[355] considered that soil and rainfall analyses were insufficient for predicting sodium need. Gallagher[124] found that despite the sodium supplied in rainfall, the element was still needed as a fertiliser in Ireland.

Conclusions

It is essential that sugar beet is adequately supplied with both potassium and sodium for maximum yield. The crop responds to potassium applications in most regions of the world for few soils can adequately supply the large uptake (1·5–2·0 cwt/acre K_2O) without fertiliser. In NW Europe and areas with similar climates, soils are leached by winter rain and exchangeable soil sodium concentrations are usually less than 50 ppm; applications of sodium are also needed on such soils for maximum yield. However, as the functions of potassium and sodium in the plant are similar, applications of one element can largely replace a shortage of the other. Thus it is sometimes more convenient to increase the potassium dressing and give no sodium. In Great Britain, where sodium costs much less than potassium, it is economically better to give sodium

TABLE 37

SUMMARY OF OPTIMUM POTASSIUM FERTILISER DRESSINGS IN RELATION TO EXCHANGEABLE SOIL POTASSIUM

Exchangeable soil K (ppm)	ADAS index	Without sodium fertiliser and soil Na <50 ppm		With sodium fertiliser and soil Na >50 ppm		Organic soils	
		(cwt/ acre K_2O)	(kg/ha K)	(cwt/ acre K_2O)	(kg/ha K)	(cwt/ acre K_2O)	(kg/ha K)
0–60	0	2·5	260	2·0	210	2·5	260
61–120	1	2·0	210	1·0	105	1·5	160
121–240	2	1·0	105	0·5	55	0·8	85
241–400	3	0·5	55	0	0	0·4	40
401–600	4	0	0	0	0	0	0

on most soils, although an application of both potassium and sodium ensures maximum yield.

Table 37 shows optimum potassium fertiliser dressings in relation to the concentration of ammonium nitrate exchangeable soil potassium. Amounts of potassium fertiliser required for maximum yield on fields not treated with sodium fertiliser and containing less than 50 ppm Na are much greater than the amounts needed when sodium is applied or where soil contains much sodium. Organic soils also need less potassium than most mineral soils. Although

TABLE 38

SUMMARY OF OPTIMUM SODIUM FERTILISER DRESSINGS IN RELATION TO EXCHANGEABLE SOIL SODIUM

| Exchangeable soil Na (ppm) | Sodium dressing | |
	(cwt/acre NaCl)	(kg/ha Na)
5·0–10·0	4	200
10·1–15·0	3	150
15·1–20·0	2	100
20·1–50·0	1	50
>50·1	0	0

exchangeable soil potassium measurements are misleading on some fields, fertiliser applications based on exchangeable soil potassium values make more profitable use of fertiliser than a uniform application to all fields.

Table 38 shows the optimum sodium fertiliser dressing in relation to the concentration of exchangeable sodium. Such exchangeable sodium values represent practically the whole of the plant available sodium, for the element is weakly held by soil colloids. Thus it leaches readily and an application of sodium fertiliser leaves no residue in the soil for the next sugar-beet crop.

TABLE 39

SUMMARY OF OPTIMUM POTASSIUM AND SODIUM FERTILISER DRESSINGS IN RELATION TO SOIL TEXTURE

Texture	K dressing without Na (cwt/acre K_2O)	(kg/ha K)	K dressing with Na (cwt/acre K_2O)	(kg/ha K)	Na dressing with K (cwt/acre NaCl)	(kg/ha Na)
Sand	2·0	210	1·2	125	3	150
Loam	1·5	160	0·5	55	3	150
Silt	1·2	125	—	—	0	0
Clay	1·0	105	0	0	3	150
Organic	0·5	55	—	—	0	0

Soil texture greatly affects the magnitude of the response to potassium and sodium, and the amount of each needed in fertiliser for maximum yield. Table 39 summarises the dressings needed by sugar beet on five soil types. Clay soils need least fertiliser because they usually contain most available potassium. Organic soils usually contain sufficient sodium for maximum yield without giving it in fertiliser. Sodium fertiliser is also considered undesirable on silty soils where the structure may be damaged.

Chapter 5

Calcium

Calcium is an important major nutrient for sugar beet, the uptake by an average crop being greater than phosphorus or magnesium, but less than nitrogen or potassium. This is often overlooked because the crop only grows well in near-neutral or alkaline soils where, in temperate climates, the clay-complex is dominated by calcium ions. Soils where sugar beet is grown in Great Britain often contain several thousand lb/acre of exchangeable calcium, so uptake of 50 lb/acre is negligible. In contrast, exchangeable potassium is rarely more than a few hundred lb/acre and uptake of 100–200 lb/acre by the crop must be supplied in part by fertiliser. Much sugar beet in Great Britain is grown on calcareous soils which contain huge reserves of calcium in the form of carbonate, but where the crop is grown on soils which are naturally acid, losses of calcium by leaching (commonly 750 lb/acre/annum) and removal in crops *is* important and needs to be rectified periodically by liming.

Calcium as a plant nutrient

AMOUNT OF CALCIUM IN SUGAR BEET

Wallace[353] found that healthy sugar-beet leaves contained 2·65% Ca in dry matter whereas those with calcium deficiency contained 0·66%. Cooke[66] reported that a 15 ton/acre crop of sugar-beet roots removed 0·71–0·89 cwt/acre Ca. At Broom's Barn the average concentration in the dried crop at harvest is 1·00% Ca in the tops and 0·24% Ca in the roots. Table 40 shows the range of values over six years and the quantity removed in the tops and roots. Finkner et al.[121] showed that nitrogen and phosphorus fertilisers significantly decreased the concentration of calcium in fresh sugar-beet roots from 0·016 4% to 0·015 1% Ca and that different varieties contained significantly different concentrations of calcium.

CALCIUM-DEFICIENCY SYMPTOMS

Sugar beet shows characteristic symptoms (often called 'tip-burn') when the supply of calcium is short or when uptake of the element is

TABLE 40

CONCENTRATION AND QUANTITY OF CALCIUM IN SUGAR BEET AT HARVEST AT BROOM'S BARN: MEANS, 1965–70

| | Ca concentration (%) | | (kg/ha) | | Quantity of Ca in the crop | | (cwt/acre) | |
	Tops	Roots	Tops	Roots	Total	Tops	Roots	Total
Means	1·00	0·24	41	22	63	0·33	0·18	0·50
Range	0·7–1·6	0·16–0·35	18–67	12–31	30–98	0·14–0·53	0·10–0·25	0·24–0·78

suppressed by excess of another ion, as often happens on saline soils or after sea-water flooding. Ulrich and Hills[347] report that the first signs of deficiency are the crinkling and downward 'cupping' or 'hooding' of young leaves. The laminae may be nearly normal in size, with little crinkling; when the deficiency is more severe they can be reduced in size to a small area of necrotic tissue at the end of the petiole.[187] Later, the growing point is damaged and abnormal lateral shoots appear. The necrosis sometimes spreads to the root, fibrous root development is diminished and the leaves wilt as a result.

RESPONSE TO CALCIUM AS A PLANT NUTRIENT
Calcium deficiency is not normally due to an absolute shortage of calcium in the soil, but to an unbalanced nutrient supply; excess of ions such as sodium and magnesium decrease uptake of calcium. Under these circumstances, where soil is neutral or alkaline, it is conceivable that the crop might respond to calcium applications, but there are few reports of this in practice. Albrecht[11] showed a benefit from calcium sulphate on irrigated desert soils in California, USA. The organic matter concentration in the soil was only 1% and the pH about 8. Calcium sulphate increased the concentration of calcium in the crop and the yield of sugar.

Soil acidity

INJURY SYMPTOMS
Like barley, mangolds, clover and other legumes, sugar beet is very sensitive to acid soil. Being a deep-rooted plant, the subsoil as well as the plough layer must not be acid otherwise plants with stubby roots result. Acid soil injury is the most important seedling trouble of sugar beet every year in Great Britain, and more seedling failures are due to this cause than any other.[187] Germination is often satisfactory in acid soil but the seedlings grow slowly and the edges of the cotyledon leaves are often red and unusually erect. Many plants die, leaving an irregular plant stand. Plants which survive sometimes have yellow leaves, sometimes due to manganese toxicity (*see* page 86).

Besides these foliar symptoms, the roots are also affected. Many roots die in acid soil, and fresh lateral roots which develop from the primary root are fibrous and fangy. Secondary attacks by fungi are also associated with acid soil injury.[187]

SATISFACTORY RANGE OF SOIL pH FOR SUGAR BEET
Applying calcium in the form of oxide, carbonate, hydroxide or sulphate effectively neutralises soil acidity. The minimum value of

soil pH to avoid acid damage is generally agreed to be 5·3.[249,26,187] If the soil has a small available phosphorus concentration[26] or if the soil is very dry[187] this value may be considerably increased. On the other hand, sugar beet on some organic soils is not damaged at even lower pH.[101]

There is much evidence that for maximum sugar beet yield, soil pH must be considerably greater than this minimum value where obvious damage is avoided. When lime is applied the concentration of elements such as aluminium, manganese and iron is decreased. Some of these elements, particularly aluminium, are toxic to plant growth but liming to pH 5·6–6·0 removes most of the aluminium from the soil solution. There are benefits of liming to pH values greater than that required to inactivate such elements, for the availability of calcium, magnesium, phosphorus, potassium, sulphur, boron, copper and zinc is increased. Biological activity is also increased with more rapid decomposition of plant residues. Equally, over-liming is to be avoided for an excess of calcium ions may antagonise the uptake of some elements such as magnesium, boron and manganese. On mineral soils there is little evidence that this is important in practice for sugar beet grows without shortage of these elements on naturally calcareous soils of pH 8·4. However, when liming for the crop on organic soils, too much should not be given because manganese deficiency may result.

METAL TOXICITIES

Soil acidity sometimes causes excessive uptake of some metal ions, particularly manganese, resulting in a characteristic chlorosis and stunting of affected plants. The laminae and petioles are uniformly pale yellowish green and the plants have an upright appearance with the margins of the leaves rolling inwards.[150] Leaf manganese concentration is of the order of several 1 000 ppm in dry matter—more than ten times the usual value for normal plants (see Table 59). Hale et al.[150] found the values ranged from 1 250 to 3 020 ppm for affected plants. However, Brown et al.,[42] whilst investigating a previously unexplained 'bronzing' of sugar beet foliage on acid soils in California, found that sugar beet could tolerate much larger concentrations of manganese in the leaf tissue than these without any obvious ill-effect. In the field the crop responded to potassium fertiliser, which also cured the bronzing. Values of up to 5 600 ppm Mn in the dry matter were reported in apparently healthy plants.

Hewitt[176] investigated the effect of adding several 'heavy' metal ions to sugar beet in sand culture. Divalent copper, cobalt and cadmium were very toxic, causing chlorosis which looked like iron deficiency. Zinc, vanadium, chromium, manganese and lead were less toxic, in that order. The zinc application produced symptoms similar to those of manganese deficiency.

Determination of lime requirement

The amount of lime needed to increase soil pH to 7 is determined in the laboratory by titration, and is usually expressed in cwt/acre CaO (chalk and limestone—both $CaCO_3$—have about half the neutralising value of CaO and factory waste lime about one-quarter). The amount of lime which needs to be applied in the field is usually about twice this amount so it is conventional to multiply by a 'field factor' of 2.

More often, the amount of lime applied to soils being prepared for sugar beet is interpreted from pH determinations made in the field. Even small fields often contain areas which are severely acidic and it is preferable that these receive a large dressing, rather than the whole field a smaller, uniform dressing. Many growers lime sugar-beet fields to a higher pH than is strictly necessary, so decreasing the likelihood of partial crop failure due to acid soil injury.

EXPERIMENTS WITH LIME
Surprisingly few thorough investigations have been made to determine optimum pH for sugar beet. In pre-war experiments, Morley Davies[249] tested the effects of additions of lime for sugar beet on an acid sandy soil. The results in Table 41 show that 25–50 cwt/acre

TABLE 41
EFFECT OF CALCIUM CARBONATE ON YIELD OF SUGAR BEET AND ON SOIL pH
(after Morley Davies[249])

Dressing		Yield of roots			
$CaCO_3$ (cwt/acre)	Ca (t/ha)	(ton/acre)	(t/ha)	Sugar (%)	Soil pH
0	0	4·5	11·3	16·7	4·9
25	3·1	9·1	22·9	17·1	5·3
50	6·3	10·3	25·9	17·1	5·5
100	12·6	10·1	25·4	17·2	5·9

calcium carbonate was needed for maximum yield, but increasing the pH above 5·5 did not increase yield further. More recently, Schmid[304] tried to neutralise a poor acid soil to a depth of three feet in different climatic zones of Germany. Lime applications needed for 80 % efficacy were determined and their effect on yield of several crops measured. In 12 experiments with sugar beet, liming to increase pH from 6·0 to 7·3 increased root yields by nearly 16%.

McEnroe and Coulter[235] made a survey on over 3 000 farms in Eire to determine the effect of pH on sugar percentage and yield of sugar beet. Table 42 shows the relationship between pH, sugar

TABLE 42

RESULTS OF A SURVEY IN EIRE OF SOIL pH, SUGAR PERCENTAGE AND YIELD
(after McEnroe and Coulter[235])

pH	Sugar (%)	Sugar yield	
		(cwt/acre)	(t/ha)
4·5–5·9	15·2	34·2	4·29
6·0–6·5	15·4	34·6	4·34
6·6–7·0	15·6	37·6	4·72
>7·0	15·8	39·3	4·93

percentage and sugar yield. As the pH increased from less than 6·0 to greater than 7·0, the sugar percentage increased from 15·2 to 15·8 and the sugar yield from 34·2 to 39·3 cwt/acre. The authors point out that some other factor (*e.g.* soil type or good farm management) may be correlated with the large pH values, a danger with surveys of this kind. However, they still considered that the optimum pH was over 7·0 and only 34% of the farms sampled were in this category.

EFFECT OF pH CHANGES ON GROWTH OF SUGAR BEET IN GREENHOUSE STUDIES

Ulrich and Ohki[343] grew sugar-beet plants in solutions kept at pH 4·0–9·0 and found that growth was greatly affected, particularly at extremes of pH. At pH 4·0 leaves were small, dark green and decreased in number, and root development was poor; growth was also decreased at pH 9·0 as shown by the yields of tops and roots in Fig. 14. Leaf colour was affected little by change in pH. Changes in pH had little effect on the nitrogen status of the plants but increasing pH decreased phosphorus concentration throughout the range tested. However, even at pH 9·0 the plants appeared to contain sufficient phosphorus for satisfactory growth. The concentrations of potassium, sodium and calcium were affected little by changes in pH; they were certainly not decreased sufficiently to cause a deficiency. None of the major nutrients caused the poor growth at small and large pH values and the authors suggested that it was a direct result of hydrogen or hydroxyl ions on the plant metabolism.

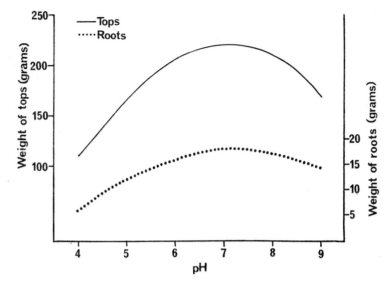

FIG. 14. Effect of culture solution pH on weight of tops and roots.[343]

Brown *et al.*[42] grew sugar beet in the greenhouse in two acid soils of pH 4·5–5·0 in California, USA, to study the effect of manganese uptake. Although plants took up large amounts of the element (up to 6·6% Mn in leaf blades) there was no apparent toxic effect. Supplying phosphorus and potassium was more important than liming for maximum yield on such soils.

Factory waste lime

Impurities in the juice extracted from sugar beet in the factory process are removed by adding calcium hydroxide; then carbon dioxide is passed through the suspension, which precipitates calcium carbonate together with the impurities. The precipitate is filtered and carried in water to settling ponds where some of the water evaporates or drains away, leaving wet calcium carbonate and the organic and mineral impurities.

Hanley[155] analysed many samples of this by-product of the sugar-beet industry over several years and showed that it contained 50% water and 40% calcium carbonate. In addition, the wet sample contained 0·3% N, 0·8% P_2O_5, 0·1% K_2O and 0·8% MgO.

Factory waste lime is therefore useful not only for correcting soil acidity but supplies considerable quantities of plant nutrients as

TABLE 43

GROUND LIMESTONE COMPARED WITH FACTORY WASTE
LIME FOR SUGAR BEET
(after Morley Davies[249])

	Year of application			Two years after application		
	Soil pH	Sugar yield (cwt/acre)	(t/ha)	Soil pH	Sugar yield (cwt/acre)	(t/ha)
No lime	5·4	33·1	4·15	5·0	13·2	1·66
Limestone	5·9	37·1	4·66	5·4	23·7	2·97
Dried factory waste lime	6·1	39·6	4·97	5·5	26·1	3·28
Wet factory waste lime	6·0	38·0	4·77	5·5	23·8	2·11

well. Mackenzie[236] found that a dressing of 10 ton/acre contained
0·50 cwt/acre N, 1·50 cwt/acre P_2O_5, 0·20 cwt/acre K_2O and 1·50
cwt/acre MgO.

The quantity of water in the waste lime is a problem as it increases
costs of haulage and makes uniform spreading difficult. Fortunately,
much of the lime can be used in sugar-beet fields, which are usually
within a short distance of the factories, and spreading is facilitated
by allowing the lime to dry in a heap in the corner of the field.

FACTORY WASTE LIME COMPARED WITH OTHER LIMING MATERIALS
Morley Davies[249] compared factory waste lime (both wet and dried)
with ground limestone. In the year of application and after two
years, the three forms of lime had a similar effect on pH of the soil
and on sugar yield (Table 43); overall factory waste lime was slightly

TABLE 44

EFFECT OF FACTORY WASTE LIME AND GROUND LIMESTONE ON
SOIL pH AND YIELD OF OATS AND BARLEY
(after Mackenzie[236])

	Dressing (ton/acre)	(t/ha)	pH	Grain yield (cwt/acre) Oats	Barley	(t/ha) Oats	Barley
No lime	0	0	4·5	35	28·4	4·39	3·56
Waste lime	30	75	5·5	+8·0	+19·7	+1·00	+2·47
Waste lime	60	150	5·7	+6·7	+21·5	+0·84	+2·69
Ground limestone	15	38	4·9	+3·5	+11·0	+0·44	+1·38
Superphosphate[a]	0	0	4·9	+3·6	+0·7	+0·45	+0·88
Ground limestone plus superphosphate[a]	15	38	4·5	+5·7	+12·2	+0·72	+0·90

[a] 3 cwt/acre P_2O_5 or 165 kg/ha P.

superior, especially when it had been dried, presumably due to the nutrients which it contained. Factory waste lime had a large residual effect on a deep coarse glacial sand at Tunstall, Suffolk.[128] In 1940, carrots failed on the untreated soil at pH 4·6 but the residues of 7·5 ton/acre waste lime (equivalent to 3 ton $CaCO_3$) applied eight years before made possible a crop of 23 ton/acre of roots. Mackenzie[236] compared factory waste lime with ground limestone and with super-phosphate on a fen soil of pH 4·5. The waste lime was more effective in increasing pH than was the limestone (Table 44) both in the surface and at depth, and also gave greater yields of oats and barley than the limestone (also shown in Table 44). As the response to phosphorus was small, Mackenzie suggested that the waste lime was more effective than the ground limestone because it was more finely divided.

Chapter 6

Magnesium

Much attention has been given recently to the magnesium nutrition of sugar beet in Britain, particularly where the crop is grown on sandy soils. Most sugar-beet crops are now adequately supplied with nitrogen, phosphorus and potassium and farmers are looking for other factors which are limiting yield. Table 45 shows that of common agricultural crops, sugar beet has a relatively large requirement of magnesium. Where intensive cropping is practised on stockless farms, magnesium is continually removed from the soil and, until recently, little returned. Sufficient exchangeable magnesium is released naturally during weathering in most soils to make up for that removed, and rain and liming materials also supply magnesium. However, on some soils these sources are insufficient to supply the needs of sugar beet—deficiency symptoms develop on the leaves and yield is decreased. Consequently, increasing quantities of magnesium-containing fertilisers are being used to correct deficiencies.

Amount of magnesium removed by sugar beet

Bolton and Penny[31] found the average amount of magnesium removed by sugar-beet tops and roots at Woburn was 20, 33 and 41 lb/acre/annum when none, 44 and 88 lb/acre magnesium respectively was given as fertiliser. Jacob[192] in a review of magnesium as a plant nutrient gave the uptake of sugar beet as 31 lb/acre. On Barnfield at Rothamsted, Warren and Johnston[360] found sugar beet removed much less—only 11 lb/acre without and 13 lb/acre with magnesium fertiliser. Adams[5] measured the uptake of magnesium by sugar beet periodically from June to October at Rothamsted; it increased rapidly until September and was then fairly constant at 24 lb/acre (assuming 30 000 plants/acre). Durrant and Draycott[110] found that maximum uptake of magnesium was 10 lb/acre (in August) but this decreased to 7 lb/acre by harvest (in November). Leaves and petioles of sugar beet contain much magnesium, so if these are left on the field removal is less. Table 45 shows the distribution between roots

TABLE 45

DRY MATTER YIELD AND QUANTITY OF MAGNESIUM REMOVED
IN CROPS AT BROOM'S BARN: MEAN OF FOUR YEARS

Crop	Yield dry matter (ton/acre)		(t/ha)		Quantity of Mg (lb/acre)			(kg/ha)		
	Grain	Straw	Grain	Straw	Grain	Straw	Total	Grain	Straw	Total
Spring barley	1·55	1·43	3·89	3·59	3·8	1·5	5·3	4·3	1·7	5·8
Winter wheat	2·24	1·36	5·63	3·42	3·3	2·7	6·0	3·7	3·0	6·7
Spring beans	1·41	1·03	3·54	2·59	2·9	2·3	5·2	3·3	2·6	5·9
	Leaves	Roots	Leaves	Roots	Leaves	Roots	Total	Leaves	Roots	Total
Potatoes	—	2·80	—	7·03	—	4·1	4·1	—	4·6	4·6
Ryegrass	1·78	—	4·47	—	4·8	—	4·8	5·4	—	5·4
Sugar beet	2·42	3·51	6·08	8·81	12·4	10·3	22·7	13·9	11·6	25·5

and tops in a long-term experiment at Broom's Barn where the crops
were healthy and yielded well. Averaging the published results, the
amount of magnesium removed from the soil was 24 lb/acre where
the whole crop was removed, about half of which was in the roots.

Magnesium deficiency symptoms

DESCRIPTION
Typical symptoms of magnesium deficiency on sugar-beet leaves have
been described by Hale et al.,[150] Wallace,[354] Björling[22] and Hull.[187]
The first sign of deficiency is the appearance of small, pale yellow areas
at the distal margins of the middle-aged leaves. Tissue in these
areas enlarges and the edge of the leaf becomes fluted. Lemon yellow
areas then spread down between the veins and appear as well-defined
lobes orientated towards the mid-rib, with distinct demarcation
between yellow and green areas due to thickening and enlargement of
affected tissue.

The chlorotic areas on each leaf are usually continuous, but
isolated patches may develop between the veins. Typically the colour
of these areas is pale yellow, distinguishing the symptoms from those
of virus yellows which are orange-yellow,[187] but individual plants
sometimes have leaves which are dark yellow or orange. Usually
within a few weeks the yellow areas become necrotic, beginning at
the edges of the leaf. The necrotic tissue is dark brown and brittle

so that it breaks away easily and the leaves have a ragged appearance. Sometimes the distal portion of a leaf may break off and the leaf becomes truncated; the remaining portion is often green with little yellowing. On other leaves, particularly late in the season, necrotic lesions develop without any preliminary yellowing. Occasionally, leaves on plants affected by magnesium deficiency become yellowed uniformly all over, with small scattered necrotic areas somewhat resembling leaf-spot diseases.

TIME OF APPEARANCE

Symptoms usually first appear in July or August in England, although in some cases symptoms are not noticed until late in September. In severe cases all leaves except the youngest eventually become affected and symptoms persist until harvest; however, plants often recover and in autumn only a ring of fully-expanded leaves may show symptoms. This probably indicates a period of shortage followed by improved supply of magnesium, an effect demonstrated by Ford[122] on magnesium-deficient apple leaves.

Magnesium deficiency symptoms also appear spasmodically early in the year on sugar-beet seedlings when they have two to eight true leaves. Dunning and Cooke[108] thought that these early symptoms resulted from root damage caused by ectoparasitic nematodes, for a damaged root system would be less efficient at taking up enough magnesium, particularly where the soil supply is short. Seedlings affected in this way usually recover during June and the symptoms completely disappear, particularly when growing conditions improve, but root damage is still evident at harvest.

ACREAGE AFFECTED

Fieldmen of the British Sugar Corporation record the acreage of sugar beet with magnesium deficiency symptoms every year.[330,86] The mean area (in acres) affected each year, for four-yearly periods, was: 1946–49—3 500; 1950–53—16 100; 1954–57—10 600; 1958–61—33 200; 1962–65—54 400; 1966–69—40 000. The area affected varies widely from year to year but these figures show a general increase during this period. The increase may be partly due to better reporting, but other factors have changed which could account for it, in particular, farming practices in the sugar beet growing areas have changed. Draycott[84] showed that there has been a large decrease in the amount of farmyard manure, which supplies magnesium, used for sugar beet during the last ten years (see page 131), presumably due to less mixed farming and more intensive cropping. Modern fertilisers contain less magnesium than previously, and the larger crops grown now need more magnesium and also remove more from the soil.

TABLE 46

PERCENTAGE OF PLANTS AFFECTED AND TIME OF FIRST RECORD-
ING OF MAGNESIUM DEFICIENCY SYMPTOMS: MEAN AREA
AFFECTED, 1960–64
(from British Sugar Corporation fieldmen's records)

Percentage of plants affected	May	June	July	August	September
			(acres)		
1–20	272	1 846	18 465	17 004	6 755
21–60	136	118	2 502	3 295	1 743
61–100	81	50	491	644	395
			(ha)		
1–20	110	748	7 478	6 887	2 736
21–60	55	48	1 013	1 334	706
61–100	32	20	199	261	160

Fieldmen also record the percentage of plants affected in each field
from May to September and Table 46 shows averages over a five-year
period. Most of the acreage affected was recorded in July and August
and less than 20 % of the sugar beet showed symptoms on the majority
of fields.

EFFECTS OF MOISTURE SUPPLY
The variation from year to year in the area affected by magnesium
deficiency symptoms is probably related to weather conditions in
spring and summer and most authors are in agreement that there is
most deficiency in dry summers. Harrod and Caldwell[161] found that
irrigated sugar beet gave less response to magnesium than when water
was withheld. Will[373] noted that magnesium deficiency was more
severe under very dry growing conditions in New Zealand. Draycott
and Durrant[85] showed that the total amount of rainfall in June,
July and August was negatively correlated with the acreage of sugar
beet affected by magnesium deficiency symptoms in each of the
years 1959–68. It seems likely from this circumstantial evidence that
magnesium is more available to sugar-beet roots when the soil is
moist, although experiments at Broom's Barn made to provide
direct evidence have so far been inconclusive.

Concentration of magnesium in sugar beet

RELATIONSHIP TO SYMPTOMS
Table 47 summarises reports of the magnesium concentration in the
dry matter of sugar-beet leaves showing symptoms of magnesium

deficiency compared with the concentration in leaves without symptoms. On average, leaves with symptoms had much less magnesium (0·130% Mg) than those without (0·498% Mg), but the range of values varied widely: 0·010–0·300% Mg for leaves with symptoms and 0·100–0·700% Mg for those without symptoms. The wide range of values is probably due to different sampling techniques (*e.g.*

TABLE 47

CONCENTRATION OF MAGNESIUM IN DRY MATTER OF SUGAR BEET LEAVES WITH AND WITHOUT MAGNESIUM DEFICIENCY SYMPTOMS

Leaf magnesium (Mg %)		References
With symptoms	*Without symptoms*	
0·150	0·390	Hale *et al.*[150]
0·096	0·546	Wallace[353]
0·170	0·490	Björling[22]
0·010–0·030	0·100–0·700	Ulrich[345]
0·096	0·546	Jacob[192]
0·149–0·219	0·217–0·559	Birch *et al.*[21]
0·110	0·480	Bolton and Penny[31]
0·100–0·200	0·200–0·650	Draycott and Durrant[83]
Range 0·010–0·219	0·100–0·700	
Means 0·120	0·444	

whether the petiole was included), different ages of plants, concentration of other ions and varietal differences, some of which are dealt with below.

CHANGE WITH AGE

As with most plant nutrients, concentrations of magnesium in sugar-beet dry matter vary considerably with the age of the plant. Durrant and Draycott[110] sampled sugar beet at monthly intervals on a magnesium-deficient field from seedling stage to harvest and found a rapid decrease at first in the concentration of magnesium in the top (leaf plus petiole), followed by a very gradual decline (Table 48). A magnesium dressing doubled the magnesium concentration but the percentage of plants with symptoms was still closely related to the magnesium concentration at all stages of growth.

In two experiments at Rothamsted described by Adams,[5] the magnesium concentration in leaves of sugar beet did not decrease rapidly during the season. In both experiments it increased at first

TABLE 48

MAGNESIUM CONCENTRATION IN SUGAR BEET TOPS AND
PERCENTAGE OF PLANTS WITH MAGNESIUM DEFICIENCY
SYMPTOMS
(after Durrant and Draycott[110])

	Without kieserite		With kieserite (5 cwt/acre or 628 kg/ha)	
Sampling date	Mg concentration (%)	Plants with symptoms (%)	Mg concentration (%)	Plants with symptoms (%)
May	0·318	0	0·780	0
June	0·283	0	0·750	0
July	0·172	5	0·476	0
August	0·125	72	0·228	0
September	0·115	95	0·198	2

and decreased in the autumn. Petiole magnesium increased slightly
from May to October in one experiment and decreased slightly in
the other; root magnesium decreased throughout in both experiments.
The Rothamsted soil is well-supplied with magnesium, whereas in the
field where a rapid decrease in leaf magnesium concentration was
reported[110] the soil magnesium was small.

EFFECT OF FERTILISERS

Applications of some fertilisers for sugar beet affect the concentration
of magnesium in the dry matter, particularly of leaves and petioles.
Hale et al.[150] showed that sodium and, to a lesser extent, potassium,
decreased magnesium concentration in the leaves and increased
magnesium deficiency symptoms; sodium also decreased the magnesium concentration even in sugar-beet leaves without deficiency

TABLE 49

EFFECT OF MAGNESIUM AND SODIUM ON MAGNESIUM
CONCENTRATION OF DRIED SUGAR BEET LEAVES AND
PERCENTAGE OF PLANTS WITH SYMPTOMS: MEAN OF 18
EXPERIMENTS
(after Draycott and Durrant[83])

	Na_0	Na_1	Na_0	Na_1
	Mg concentration (%)		Plants with symptoms (%)	
Mg_0	0·369	0·352	12·5	16·8
Mg_1	0·440	—	5·0	6·4

Mg_1—5 cwt/acre or 628 kg/ha kieserite.
Na_1—3 cwt/acre or 377 kg/ha crude sodium chloride.

symptoms. Tinker[330] confirmed the effect of sodium but found that the magnesium concentration in sugar-beet leaves was affected little by applications of nitrogen, potassium or calcium. Table 49 also shows that sodium and magnesium fertilisers changed the magnesium concentration, with corresponding changes in the percentage of plants with symptoms.

In the experiments of Hale *et al.*,[150] although sodium only decreased magnesium concentration slightly, it greatly increased deficiency symptoms and the authors suggested that the greatly increased sodium concentration in the leaves had a *direct* effect in increasing symptoms. It seems more likely that the effect of sodium applications on symptoms is only large when the magnesium concentration in the plant is small. Thus the absolute amount of magnesium in the leaves governs whether symptoms appear, and the concentration of sodium in the leaves only *indirectly* affects symptoms.

Bolton and Penny[31] found that potassium applications decreased the magnesium concentration of several crops at Woburn including sugar beet, but yield response to magnesium fertiliser was not increased, a result confirmed by Birch *et al.*[21] A more detailed account of the interactions of fertilisers on yield in magnesium-deficient fields is given on page 104.

Magnesium in soil

Cooke,[66] reviewing the magnesium economy of soil, suggested that it was not surprising that serious magnesium deficiencies are uncommon in Britain, for uptake by many crops is of the same order as the amounts replaced in fertilisers, manures, liming materials and rain-water together. Nevertheless, there is evidence that magnesium deficiency in sugar beet is becoming more widespread in certain areas, and recent reports show that the crop responds economically to magnesium fertiliser.[330,31,83,90]

The soils where magnesium deficiency symptoms are most common are usually loamy sands or sandy loams, which have less reserve of magnesium than clay soils.[298] Sugar beet has a larger than average requirement of magnesium and in East Anglia (where the British crop is most concentrated) the local chalk used as liming material contains much less magnesium than dolomitic limestone used in some sugar-beet growing areas. These factors, together with the decreased use of organic manures, all accentuate the inability of some soils to supply magnesium from reserves. Magnesium fertilisers appear to be relatively inefficient (compare nitrogen, page 9) for 90 lb/acre Mg only increases uptake by 4 lb/acre.[110]

Magnesium in rain, limes and organic manures

Rainfall in Great Britain brings 1 to 10 lb/acre/annum because the concentration of Mg in rain varies between 0·1 and 0·5 lb Mg/acre/inch. Chalk contains about 10 lb Mg/ton so routine applications of a few tons/acre make important additions to the soil. Dolomitic limestone is a much more concentrated source of magnesium, one ton supplying about 250 lb Mg. Factory waste lime is discussed on page 89. Farmyard manure contains 2–3 lb Mg/ton, and straw 1–2 lb/ton. A 10 ton/acre dressing of farmyard manure would therefore supply 20–30 lb/acre, which would be sufficient for most crops, but the straw remaining from a cereal crop would only leave about 3 lb/acre.

Predicting response to magnesium by soil analysis

Many soils contain large reserves of magnesium in unweathered minerals and clay, but magnesium which is available to plants is either in solution or exchangeable. Attempts have therefore been made to extract 'available' magnesium by leaching or equilibrating soil with aqueous solutions, usually of ammonium salts. Birch et al.[21] tested response by sugar beet to magnesium fertiliser on soils ranging from 32–274 ppm Mg (extracted with N ammonium acetate). Although plants on some fields had symptoms of deficiency, yields were not increased significantly in any experiment. However, when Harrod and Caldwell[161] tested response in five experiments on soils ranging from 20–35 ppm Mg (extracted by leaching with N ammonium acetate), in three of the five fields there were substantial increases in yield of sugar. All three responsive soils had 30 ppm Mg or less, suggesting that the reason why Birch et al.[21] found no relationship between response and soil analysis was because all their soils had sufficient available magnesium for near-maximum yield.

Bolton and Penny[31] tested response to magnesium by a range of crops at Woburn and found that sugar beet responded when the exchangeable soil magnesium (extracted by leaching with N ammonium acetate, soil:extractant ratio 1:15) was less than 2 mg Mg/100 g soil (20 ppm). They suggested that a soil value of 4 mg Mg/100 g was sufficient for most crops. Tinker[330] measured percentage response to magnesium sulphate in seventeen experiments on fields where response was expected. The soils were shaken with 0·01 M calcium chloride or leached with 1 N ammonium acetate and the magnesium measured in the extracts. He found that the calcium chloride values were linearly related ($r = -0·64$) to response but

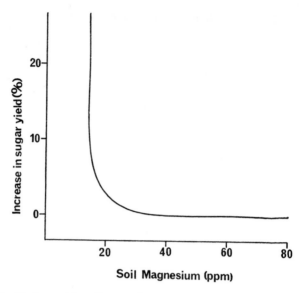

Fɪɢ. 15. Exchangeable soil magnesium and the percentage increase in sugar yield from magnesium fertiliser.[90]

that the ammonium acetate values were not (although they were all small).

More recently, Draycott and Durrant[90] suggested a non-linear relationship as shown in Fig. 15. They measured response in 53 experiments on soils with a wide range of soil Mg values and also compared four methods of extraction (Table 50). Although the amounts of magnesium extracted by the four methods differed, they

TABLE 50

EXCHANGEABLE MAGNESIUM IN SOILS FROM FIELDS WHERE SUGAR BEET YIELD WAS INCREASED BY >5% AND WHERE YIELD WAS NOT INCREASED BY MAGNESIUM FERTILISER EXTRACTED IN FOUR WAYS

(after Draycott and Durrant[90])

	Soil: Extractant	No increase in yield (ppm Mg)	Increase in yield >5% (ppm Mg)
N ammonium acetate leaching	1:10	>50	<20
N ammonium acetate shaking	1:10	>50	<25
N ammonium nitrate shaking	1:5	>35	<20
0·01 M calcium chloride	1:2	>20	<10

were significantly correlated to each other. All four methods predicted fields where sugar beet gave a large response and those which did not respond. In the intermediate range no method of analysis predicted the magnitude of response with any certainty, largely due to experimental error in measuring small responses. Nevertheless, soil analysis was of value in distinguishing fields where yield loss from magnesium deficiency was serious.

RELATIONSHIP BETWEEN AVAILABLE SOIL MAGNESIUM, DEFICIENCY SYMPTOMS AND PLANT MAGNESIUM

Björling[22] showed that magnesium deficiency symptoms in sugar beet were related to available soil magnesium. He analysed soil samples from fields where plants had symptoms and also from fields where sugar beet were healthy. Magnesium was extracted from the soils by Morgan's solution (0·5 M acetic acid, 0·74 M sodium acetate at pH 4·9) and with N ammonium acetate (soil:extractant ratio 1:5). Deficient soils contained 49 ppm Mg (Morgan) and 41 ppm (ammonium acetate); where plants were healthy the soil contained 127 ppm (Morgan) and 117 (ammonium acetate). Morgan's solution consistently extracted more magnesium than ammonium acetate, probably because it dissolved some magnesium carbonate. As most soils where sugar beet is grown are neutral or alkaline, acidic extractants for magnesium over-estimate the available fraction. Similarly, Warren and Johnston[360] showed that regular applications of sulphate of ammonia were sufficient to dissolve and make available measurable amounts of magnesium for crops on Barnfield because the local chalk contained 0·4 % Mg.

Draycott and Durrant[90] showed that soil magnesium was closely related to the concentration of magnesium in the dry matter of sugar beet sampled in late summer (Fig. 16) and that the percentage of sugar-beet plants with deficiency symptoms was related to soil magnesium. These are good indications that ammonium salt solutions do extract magnesium available to sugar beet.

EFFECT OF OTHER IONS

Some reports of responses to magnesium, particularly with horticultural crops, have stressed that soil magnesium alone should not be used to predict requirement of magnesium fertiliser, but that the concentration of other cations (*e.g.* potassium) should also be taken into account. Fertiliser applications of cations do decrease the concentration of magnesium in sugar beet[150,31,330,83] and increase the severity of symptoms.[150,83] However, with normal applications of potassium and sodium for sugar beet there is little evidence that they affect availability of magnesium to any practical extent. Applying

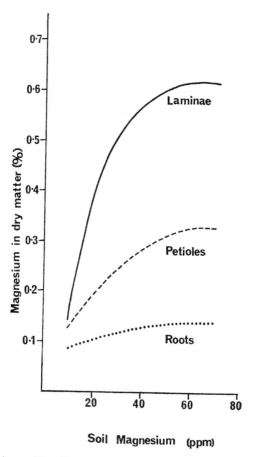

FIG. 16.   Exchangeable soil magnesium and the concentration of magnesium in dried laminae, petioles and roots.[90]

potassium decreased magnesium in sugar beet at Woburn[31] and might have been expected to increase yield response to magnesium, but did not. Draycott and Durrant[83] in experiments with potassium, sodium and magnesium found that largest yields came from plots with a full dressing of all three cations.

Effects of magnesium fertilisers on yield of sugar beet

The effects of magnesium fertilisers on yield of sugar beet have been tested on relatively few fields compared with nitrogen, phosphorus and potassium fertilisers. Birch *et al.*[21] used kieserite ($MgSO_4 \cdot H_2O$ —17% Mg) on 14 fields where the soil was predominantly sandy and found that sugar yield was increased slightly but not significantly in five fields. However, when Tinker[330] tested magnesium sulphate on 17 fields where deficiency symptoms were expected by farmers, the mean increase in yield was 3·7 cwt/acre sugar (of considerable economic significance), with a small increase in yield of tops. Also, Harrod and Caldwell[161] used kieserite on five fields with small exchangeable soil magnesium concentrations and found increases ranging from 3 to 9 cwt/acre sugar on three fields. Bolton and Penny[31] reported an increase in dry matter yield of sugar beet at Woburn of 4·8% averaged over two years. Draycott and Durrant[83] made nineteen experiments on fields where previous sugar-beet crops had magnesium deficiency symptoms, and the average increase from kieserite was 3·1 cwt/acre sugar (about 6%) although on some fields increases were up to 20%.

The size of the response to magnesium varies greatly from field to field but it is related to the percentage of plants with deficiency symptoms, the concentration of magnesium in the dry matter of the plants, and to the amount of exchangeable magnesium in the soil. These are all guides to where magnesium fertiliser is needed, but the size of the response changes very rapidly from no response to a large response over a narrow range of values of soil and plant magnesium.

Thus magnesium dressings given for crops on fields where symptoms are expected are an insurance against a large depression in yield. From results of the experiments quoted, in order to avoid loss, magnesium applications would be needed to increase the magnesium status of soils to 40 ppm (90 lb/acre Mg is equivalent to 30 ppm in the top 9 in, which should suffice to increase all soils to at least 40 ppm). The returns in sugar beet on many fields would be small but large yield losses would be avoided and other crops in the rotation might benefit.

Comparisons of forms and methods of application of magnesium fertilisers

FORMS
Most experiments with sugar beet have been made with Epsom salts ($MgSO_4 \cdot 7H_2O$—10% Mg) or with kieserite. Both seem equally effective for sugar beet, no doubt because they dissolve easily in soil moisture and the magnesium is readily available to roots. Draycott and Durrant[83] compared kieserite with kainit (KCl, NaCl, $MgSO_4$—4·5% Mg) and magnesian (dolomitic) limestone

($MgCO_3$, $CaCO_3$—11% Mg). The magnesium in the kainit increased yield and plant magnesium by the same amount as an equivalent dressing of magnesium in kieserite, but the magnesium in the limestone was not nearly as effective as magnesium in kieserite. Soils prepared for sugar beet must be neutral or alkaline, so presumably magnesium in the carbonate (which is relatively insoluble) remained largely unchanged and little magnesium was immediately available to the sugar-beet roots. If used for correction of acidity on magnesium-deficient fields, magnesium limestone is an inexpensive and effective source of magnesium, with lasting qualities superior to the very soluble forms. Calcined magnesite (MgO—55% Mg) is the most concentrated form of magnesium fertiliser in use at present, some forms of which are somewhat less efficient than kieserite for sugar beet, but more experiments are needed to compare the short and long-term effectiveness of different forms of magnesium.

FOLIAR SPRAYS

Some early experiments tested the effect of correction of deficiency symptoms on sugar-beet leaves, using foliar sprays of soluble magnesium salts, usually Epsom salts, which removed deficiency symptoms from the leaves,[150] but their effects on yield were not studied. Treatment by foliar sprays *after* symptoms appear would be convenient in practice, but there is no published evidence that spraying increases yield to the same extent as applying magnesium to the soil before sowing. Tinker[330] described two experiments testing none, 0·75 and 2·00 cwt/acre Epsom salts sprayed and 2·00 cwt/acre kieserite applied dry as a top-dressing in mid-July on fields where 80–100% of the plants showed symptoms at the time. The symptoms were ameliorated but yields were not significantly increased in late October. More experiments are needed comparing seedbed treatments which are known to be effective, with foliar sprays before the spraying can be advised as a useful corrective measure.

Interactions between magnesium and other fertilisers on sugar beet yield

Tinker[330] described experiments on seventeen fields where magnesium fertiliser was tested with potassium, sodium and nitrogen. There was a small negative interaction in sugar yield between magnesium and potassium but none between magnesium and sodium. Nitrogen masked the symptoms of magnesium deficiency and there was a negative interaction between magnesium and nitrogen in sugar yield. Draycott and Durrant[83] made nineteen experiments testing combinations of magnesium, nitrogen and sodium and found no significant interactions on average between magnesium and the other two fertilisers, although all had large effects on composition of the crop and on the percentage of plants with symptoms.

As in Tinker's experiments, nitrogen decreased the number of plants with symptoms but magnesium fertiliser nonetheless increased yield by the same amount as when nitrogen was not given. Similarly, sodium fertiliser increased symptoms, but both sodium and magnesium fertiliser were needed for maximum yield. In a long-term experiment at Woburn, Bolton and Penny[31] found that potassium decreased the magnesium concentration in sugar beet and other crops but giving magnesium did not increase yield. This may have been because the concentration of exchangeable soil potassium was small and the response to potassium fertiliser was much larger (and more variable) than the small extra yield expected from magnesium. Until more detailed experiments have been done to examine interactions between magnesium and other fertilisers, the available evidence suggests that interactions can be largely ignored at present in farming practice.

Time and periodicity of application of magnesium fertilisers

In most experiments testing the effects of magnesium fertilisers on sugar beet the dressings have been applied in spring, either on the ploughed land or at some stage during seedbed preparation. There have been no experiments comparing autumn or winter with spring applications. Preliminary results[93] of some long-term experiments on deficient fields showed that kieserite was more effective when applied several years before the sugar-beet crop than when applied in the sugar-beet seedbed, indicating that time of application needs more investigation. These results also show that magnesium fertilisers are effective for a period of years, as shown for other crops by Charlesworth,[54] Greenham[136] and Allen.[12]

In the experiments of Bolton and Penny,[31] repeated annual magnesium applications gradually increased exchangeable soil magnesium by an amount roughly equal to the amount remaining after a deduction for that removed in the crops, and there was little evidence that non-exchangeable magnesium was released during six years of cropping. The implication is that the nutrient balance sheet approach (described by Cooke[66]) probably works well for magnesium, and may be of value in farming practice.

Conclusions

Experiments during the last ten years provide indisputable evidence that the yield of some sugar-beet crops can be increased economically by magnesium fertiliser. However, experiments have also established that the majority of sugar-beet crops do not respond. Nitrogen, phosphorus, potassium and sodium fertilisers give large or moderate increases in yield on most fields and marginal increases on nearly all the rest; crops do not respond to these four elements on only a few fertile fields. Response to magnesium fertiliser appears to be quite different, for yield is increased only on fields where the crop would otherwise develop deficiency symptoms.

It is estimated that at present in Great Britain less than 5% of the sugar-beet area gives an immediate profitable return from magnesium fertiliser; a further 5% probably gives a marginal profit. However, there is evidence that the area affected by deficiency has been increasing, presumably due to specialisation in crops which are sold off the farm. Where only pure chemical fertilisers and lime containing little magnesium are used, and where there are no animal residues, crops deplete soil reserves of magnesium. Most soils release enough magnesium from clay to make up the loss but some sandy soils cannot. Thus there may be justification for applying magnesium fertiliser to prevent deficiency developing. This is the subject of much research at present.

There is no infallible method of predicting where sugar beet will respond to magnesium fertiliser. The best guide is the previous sugar-beet crop—if many plants showed deficiency symptoms then a profitable return from fertiliser is likely. Crops on fields where symptoms have never been seen are unlikely to respond.

Soil analysis for available magnesium is also useful. Figure 15 shows that crops on fields with more than 50 ppm exchangeable magnesium have an ample supply from the soil. When the exchangeable magnesium is less than 25 ppm a large increase in yield is likely, but in the range 25–50 ppm crops on only a few fields respond. Although more experiments are needed to improve the precision of this soil test, it is a valuable method for distinguishing the few fields where crops are likely to respond from the majority, where magnesium fertiliser gives no return.

Chapter 7

Micronutrients or Trace Elements

In common with other plants, in addition to the major nutrient elements described in previous chapters, sugar beet needs small amounts of other nutrients. These micronutrients or 'trace elements' essential for plants are boron, chlorine, cobalt, copper, iron, manganese, molybdenum and zinc. Other elements such as germanium, nickel, rubidium, selenium and vanadium are recognised as being essential for some plants but their importance in sugar-beet nutrition has yet to be investigated.

Table 51 shows the concentration of micronutrients in sugar-beet tops and roots and the quantity in an average crop at harvest. Most

TABLE 51

CONCENTRATION AND QUANTITY OF MICRONUTRIENTS IN SUGAR BEET

| | Concentration in dry matter (ppm) | | Quantity in crop (tops plus roots) | |
	Tops	Roots	(lb/acre)	(g/ha)
Boron	40	15	0·30	335
Chlorine	2 000	1 000	17·00	19 000
Copper	7	1	0·04	44
Iron	200	100	1·70	1 900
Manganese	50	30	0·46	520
Molybdenum	7	5	0·07	80
Zinc	20	10	0·17	190

soils supply the needs of the crop from reserves, weathering minerals, rainfall, limes, fertilisers and organic manures. One dressing of farmyard manure supplies more of most micronutrients than sugar beet removes (Table 52).

On soils where natural supplies of some micronutrients are small and where farming practice depletes soil reserves, some elements

107

need to be applied for sugar beet to yield fully. In most countries where the crop is grown, boron and manganese are the only two of widespread importance. Sugar-beet crops in localised areas show sporadic symptoms of iron deficiency and some are reported to be short of copper and zinc, but these deficiencies appear to be of little

TABLE 52

MICRONUTRIENTS IN FARMYARD MANURE

| | Concentration in dry material (ppm) | | | | Amounts in 10 ton/ acre (26 t/ha) dressing | |
	Stojkovska and Cooke[322]	Heming- way[167]	Atkinson et al.[14]	Means	(lb/acre)	(g/ha)
Boron	20	84	20	41	0·37	426
Cobalt	6	1·7	1	3	0·03	28
Copper	62	20	16	33	0·30	312
Manganese	410	182	201	264	2·40	2 500
Molybdenum	—	2·3	2·1	2	0·02	19
Nickel	10	—	—	10	0·09	94
Zinc	120	—	—	120	1·10	1 123

economic importance at present. However, as yields increase and farming practices change, applications of other micronutrients may be needed in future.

BORON

Deficiency symptoms

ON THE ROOT CROP

Shortage of boron causes typical symptoms to appear not only on the leaves (as with most element deficiencies) but also on the petioles, crowns and roots of sugar beet. Brandenburg[39] first showed that boron deficiency was the cause of 'heart rot' and 'dry rot'. Heart rot is the term applied when the growing point becomes blackened and dies; dry rot describes the symptoms on the tap root which usually appear subsequently. Boron deficiency not only decreases yield but damages the roots, decreasing their value and keeping quality.

When the supply of boron to sugar beet is insufficient for normal growth, the first sign of deficiency is the death of the growing point and the youngest leaves. As the deficiency becomes more serious, the older leaves have a silvery appearance and are often covered with a network of fine cracks and sometimes show interveinal chlorosis.

Occasionally the grooves in the petioles of affected plants have transverse cracks (laddering) and the tissue is brown and cork-like. These leaves are usually more or less prostrate. Boron deficiency seems to cause unequal growth of the two faces of the leaves, resulting in torsion of the petioles and veins and, finally, in abnormal leaves.

As affected plants age, many of the leaves die and fall off. If the growing conditions improve, plants usually begin to grow again from secondary growing points. Tissue round the shoulder of the tap root also begins to decay in the advanced stages of deficiency. Such roots are invaded by fungi and begin to rot. Hull[187] has described this secondary attack and some of the fungi involved.

ON THE SEED CROP

When seed plants are boron-deficient they show pronounced symptoms. The main flowering stem is stunted and the growing point dies. Similarly, if other stems shoot, although they are taller than the first one, their growing points also die. The growing points of laterals along these stems also give rise to stunted growths which appear as small rosettes of discoloured bracts and eventually these growing points also die.[187] Brenchley and Watson[41] demonstrated the importance of boron for the sugar-beet seed crop.

DISTRIBUTION

Boron deficiency is wide-spread throughout the sugar-beet growing regions of the world. Symptoms described above from observations on the British crop are typical of those described in other parts of the world (e.g. Brandenburg[39] in Germany; Kotila and Coons[204] in USA; Lachowski[211] in Poland).

ANATOMICAL EFFECTS AND DISTRIBUTION IN THE PLANT

Rowe[294] made a detailed anatomical study of the effects of boron deficiency on sugar beet. She found that the apical meristem of the shoot, together with the youngest leaves and the newly-developed cambia, were most sensitive to boron deficiency and these were the first to degenerate. Cells of the vascular rings in the process of differentiating and sporadic groups of parenchyma cells adjacent to conducting elements were also sensitive to deficiency. Later stages of the deficiency were characterised by decay of the cambial cells and adjacent parenchyma, together with complete disintegration of the phloem. The root tip did not degenerate but merely ceased to grow. A concentration of 0·17 ppm in the culture solution was enough for normal growth and development. Recovery in boron-starved plants involved the activation of axillary buds at the top of the beet, each of which developed its own system of secondary vascular rings.

Ráb[283] found that the concentration of boron in sugar-beet roots varied greatly according to age. In young plants it was largely near the crown and root tip whereas at harvest it was concentrated in the central portion. Boron present in the leaf blade decreased greatly towards the end of the growing season, while that in the petioles changed little.

INFLUENCE OF SOIL AND WEATHER

Although most soils contain sufficient plant-available boron for healthy crop growth, the concentration in some is too small to satisfy the needs of plants with a large requirement, and deficiency symptoms appear. Plants differ widely in their ability to absorb boron from soils (and from solutions) and in their requirements of the element. Sugar beet has one of the largest requirements of boron of any common crop.

Sugar beet on light-textured soils is most prone to deficiency, particularly when alkaline, and increasing the pH of such soils by liming increases the severity of boron deficiency. Boron occurs naturally in soils mainly as the mineral tourmaline and in the organic matter, both of which contribute slowly to the water-soluble or plant-available fraction. Rainfall during winter readily leaches boron from soil and wet winters increase the prevalence of deficiency in the sugar-beet crop. Boron deficiency is also most common in dry summers, particularly when a mild, wet spring is followed by drought. Under these conditions the plants grow rapidly at first and absorb much boron; with the dry weather, the availability of soil boron decreases and deficiency symptoms appear, despite a large concentration of boron in the old leaves. Sugar-beet plants do not appear to be able to translocate this boron from old leaves to other parts of the plant, so the growing point and younger leaves may die from shortage of boron during a dry spell whilst a large amount is immobilised in the old leaves.

Concentration of boron in sugar beet

Table 53 shows the boron concentration in plants showing deficiency and in similar ones without symptoms. The concentration in the roots seems to be a very poor guide to the boron requirement. However, the concentration in sugar-beet leaves is closely related to the symptoms of deficiency. Generally, symptoms do not appear

TABLE 53

BORON DEFICIENCY AND CONCENTRATION OF THE
ELEMENT IN THE DRY MATTER

Stage of growth	Plant part	Boron concentration (ppm B) Showing deficiency symptoms	Without symptoms	Country	References
—	Leaf	4–28	25–52	Germany	Brandenburg[40]
	Root	13–14	13–15		
Aug/Sept	Leaf blade	19–35	20–426	California,	Eaton[113]
	Root	6–30	2–28	USA	
	Whole plant	13–33	10–139		
—	Leaf	6–13	10–44	Great Britain	Hale[149]
June		13–29	22–73		
August	Old leaf	10–37	26–75		
November		16–36	26–52	Great Britain	Hamence and
June		25–37	29–90		Oram[153]
August	Young leaf	8–22	25–40		
November		19–24	15–40		
—	Leaf	17	29	Great Britain	Wallace[353]
Range	Leaf	4–37	10–90		
	Root	6–14	2–28		
Means	Leaf	20	40		
	Root	13	15		

where the leaf concentration is greater than 30 ppm. Symptoms are usually present when the concentration falls below 20 ppm.

Response to boron applications

Hanley and Mann[154] made some of the first experiments in England. 0, 4, 14 and 28 lb/acre borax (sodium borate, $Na_2B_4O_7 \cdot 10H_2O$— 11 % B) was applied to the soil three weeks before sowing. Deficiency symptoms appeared in 50 % of plants in the untreated area but was decreased to 6 % in plots with 14 lb/acre borax. For maximum yield the authors recommended 20 lb/acre borax.

Hamence and Oram[153] applied 20 lb/acre borax on one of three occasions—in the seedbed, in mid-June or in late August. Liquid and solid applications were compared in each case and the experiments were made on twelve sandy fields in East Anglia, selected as far as possible as having a previous history of boron deficiency.

Yields of roots and sugar (Table 54) were increased significantly when the crop was severely deficient. Sugar percentage was also improved on these fields, but not juice purity. There were no consistent differences between liquid or solid applications whether to seedbed or to the growing crop, but August treatment never gave good results. They recommended that a routine application of

TABLE 54

RESPONSE TO BORON APPLIED IN VARIOUS WAYS
(after Hamence and Oram[153])

Year	Number of experiments	No boron	Solid, seedbed	Liquid, seedbed	Solid, June	Foliar, June	Solid, August	Foliar, August
				Sugar yield				
				(cwt/acre)				
1959	3	50·2	—	—	54·4	53·0	—	—
1960	3	62·1	64·4	65·4	63·2	63·6	—	—
1961	6	51·7	53·8	56·0	55·9	55·0	51·3	50·8
				(t/ha)				
1959	3	6·30	—	—	6·83	6·65	—	—
1960	3	7·79	8·08	8·21	7·93	7·98	—	—
1961	6	6·49	6·75	7·03	7·02	6·90	6·44	6·38

boron be made to all sugar-beet crops on light soils either as a seedbed dressing or in June as a foliar spray.

Hull[190] in four experiments (1936–43) on sandy soils showed that 20 lb/acre borax was adequate to prevent boron deficiency. Yield was increased by 2 cwt/acre sugar. Four further experiments (1945–49) gave a similar result; in these experiments the interactions between boron, nitrogen and sodium fertilisers and time of sowing (either the first week in April or the first week in May) were also investigated. Table 55 shows the results. There was a small positive interaction between boron and nitrogen for when 3 cwt/acre ammonium sulphate was applied, 20 lb/acre borax in the seedbed increased sugar yield by 0·8 cwt/acre but when 6 cwt/acre ammonium sulphate was used, yield increased by 2·0 cwt/acre. There was a small positive interaction of similar magnitude between boron and sodium chloride. Response to boron decreased with late sowing. Crops sown in April gave an extra 2·4 cwt/acre sugar whereas those sown in May gave only an extra 0·4 cwt/acre.

Berger[19] in Wisconsin, USA, measured response to borax by sugar beet in a series of experiments. The borax was either broadcast

TABLE 55

EFFECT OF NITROGEN AND SODIUM FERTILISERS AND SOWING
DATE ON RESPONSE TO BORON: MEANS OF 4 EXPERIMENTS,
1945–49
(after Hull[190])

	Sugar yield					
	Ammonium sulphate 3 6 (cwt/acre)		Sodium chloride 0 5 (cwt/acre)		Sowing date April May	
No boron	29·9	29·0	29·0	30·0	32·5	26·4⎫
Sodium borate (20 lb/acre)	30·7	31·0	29·7	32·0	34·9	26·8⎭ cwt/acre
	Ammonium sulphate 377 753 (kg/ha)		Sodium chloride 0 628 (kg/ha)		Sowing date April May	
No boron	3·75	3·64	3·64	3·77	4·08	3·31⎫
Sodium borate (22 kg/ha)	3·85	3·89	3·73	4·02	4·38	3·36⎭ t/ha

at 25 lb/acre on the ploughing and disced in, or at 15 lb/acre along-
side the seed row in two bands about 1½ in to the side of the seed
and slightly below it. The two methods of application were equally
effective. There was an average response of about 2 tons/acre roots
to the borax (Table 56) on soils containing 1·4 lb/acre available

TABLE 56

RESPONSE TO BORON IN WISCONSIN
(after Berger[19])

Available boron in soil	*Root yield*		
	No boron	Sodium borate broadcast (25 lb/acre)	Sodium borate side-dressed (15 lb/acre)
0·5 ppm	15·4	17·3	17·2⎫
0·7 ppm	14·3	14·3	14·4⎭ ton/acre
	No boron	Sodium borate broadcast (28 kg/ha)	Sodium borate side-dressed (17 kg/ha)
0·5 ppm	38·7	43·4	43·2⎫
0·7 ppm	35·9	35·9	36·2⎭ t/ha

TABLE 57

AMOUNTS AND METHODS OF APPLICATION OF BORON IN SOME SUGAR-BEET PRODUCING COUNTRIES
(after Shorrocks[310])

Country	Fertiliser with which boron is commonly incorporated	B (%)	Amount of B applied (lb/acre)	Amount of B applied (kg/ha)	Other forms of B used
Austria	Calcium ammonium nitrate	0·4	1·8	2·0	Sodium borate foliar spray
Belgium	NPK compound	0·15	2·0	2·3	Sodium borate
Denmark	PK compound	0·3	—	—	Imported fertilisers and sodium borate
Eire	NPK compound	0·22	2·5	2·8	All crops treated with the NPK compound
Finland	NPK compound	0·16	1·8	2·0	Sodium borate
France	NPK compounds	—	—	—	Up to 36 lb/acre (40 kg/ha) sodium borate used on some fields with high pH
Germany	NPK compounds	0·2	1·5	1·7	—
Japan	NPK compound	0·1	1·1	1·2	—
Iran	—	—	2·2	2·5	Applied as sodium borate
Morocco	Range of fertilisers	0·3–1·2	3·1	3·5	All crop treated with up to 5 lb/acre (6 kg/ha) B
Spain	Range of fertilisers	0·2	2·9	3·3	Sodium borate
Sweden	PK	0·3	1·3	1·5	—
Great Britain	20:10:10 ($N:P_2O_5:K_2O$)	0·3	2·1	2·4	Sodium borate
United States of America	Range of fertilisers	—	—	—	Sodium borate at 9–18 lb/acre (10–20 kg/ha)

boron on average (0·5 ppm) and no response where the soils contained 2·2 lb/acre (0·7 ppm). Lachowski[211] tested boron for sugar
beet in Poland in over 100 experiments. Yields were slightly improved
but the processing value was not affected.

AMOUNT AND METHOD OF BORON APPLICATION
Table 57 shows the amounts and methods of application of boron in
some sugar-producing countries. Boron is commonly incorporated
in compound fertilisers containing nitrogen, phosphorus and
potassium. The concentration of boron in these ranges from 0·1 to
0·4% so that a suitable dressing of the fertiliser for sugar beet
supplies 1–4 lb/acre B. Generally, the hotter the climate and the
more alkaline the soil reaction, the larger the dressing of boron.
For example, in Morocco where the growing season is hot and
dry and the soils are calcareous, the dressing is 4 lb/acre and all
sugar-beet crops there are treated.

An alternative in many countries to this method of boron application is a foliar spray of either borax (11% B) at 9–35 lb/acre or
'Solubor' (21% B) at 6–9 lb/acre. This method is convenient where
insecticidal sprays are used in the sugar-beet crop during May–July.
This treatment often costs less than seedbed dressing of boron
incorporated in the sugar-beet fertiliser.

**Amount of boron removed from soil by sugar beet and residual effects
of boron applications**

Some crops are sensitive to an exceesive supply of boron (whether
from newly applied boron fertiliser or from residues in the soil);
barley often shows boron toxicity symptoms where boron fertiliser
has been used by mistake. Consequently there have been several
investigations to determine the residue from boron applications
given to sugar beet. According to Brandenburg[40] a normal crop
of sugar beet extracts 0·27–0·36 lb/acre B. Thus of the 18 lb/acre
borax usually applied for sugar beet, 14 lb/acre remain in the soil.
Leaching experiments made by Krügel et al.[206] in Germany showed
that 75% of this was leached out of the soil during the first winter,
and by the third year after application little remained. In the climate
of Western Europe, they concluded that there was no evidence that
boron accumulated in the soil.

Very few long-term experiments have measured the effects of
continued dressings of micronutrients on crops and soils, but one
field experiment in Norway[258] continued for 23 years. Annual
dressings of two ounces of boron/acre/annum prevented boron

deficiency, and one ounce was partly successful. Giving 1 lb boron/ acre/annum injured some of the crops grown after a few years, especially on unlimed soil. Hamence and Oram[153] and Van Luit and Smilde[349] in England and Holland respectively have investigated the fate of boron applied for sugar beet. Both reports indicate that little of the boron remains two years after application, presumably because it is subject to rapid leaching during winter.

Prediction of boron requirement by plant analysis

Hamence and Oram[153] found that analysis of leaf blades of older leaves in June when it was not too late to take the necessary corrective action, was the best way of predicting where boron was needed. Symptoms appeared in all crops where the boron concentration of old leaves *at some period* was below 25 ppm, even if the symptoms did not persist throughout the season.

Prediction of boron requirements by soil analysis

Smilde[315] has recently reported on the value of soil analysis for predicting where sugar beet needs boron in Holland. Results of plant analyses from field and pot experiments showed that there was a close relationship between hot-water-soluble soil boron and leaf boron with sandy soils varying widely in pH (3·8–6·5 in potassium chloride solution) and organic matter content (1·9–9%), as shown in Fig. 17.

Soil and leaf boron concentrations were closely related to the percentage of affected plants (Fig. 18). No heart rot appeared when the hot-water-soluble boron concentration was greater than 0·35–0·40 ppm or when the leaf boron concentration was greater than 35–40 ppm. There was no loss in yield of tops, roots or sugar from boron deficiency when the soil boron exceeded 0·30–0·35 ppm.

Leaf boron decreased and heart rot increased as soil pH and organic matter increased. For soils with 0·30–0·35, 0·20–0·29, and less than 0·20 ppm B, the best dressings were 4, 10 and 15 lb/acre borax (11% B) respectively. Borax markedly increased sugar percentage of sugar beet growing in deficient soil in pots but the effect was much less pronounced in the field. Hamence and Oram[153] met with less success in predicting boron deficiency from hot-water-soluble (or from total) soil boron. They found soil analysis was an unreliable means of diagnosing boron requirement.

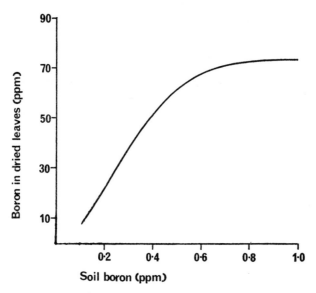

FIG. 17.   Hot-water-soluble soil boron and the concentration of boron in dried leaves.[315]

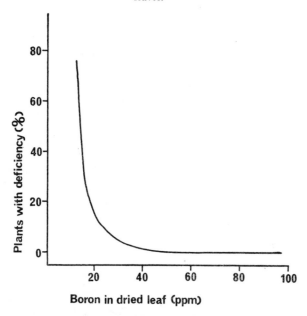

FIG. 18.   Concentration of boron in dried leaves and the percentage of plants with deficiency symptoms.[315]

CHLORINE

Chlorine is the latest element to be established and generally accepted as a micronutrient essential for plant life, but little is known about the element in relation to sugar-beet nutrition. Deficiency symptoms appear as an interveinal chlorosis on the middle-aged leaves, somewhat similar to those of manganese deficiency.[347] Deficient leaves contained 0·01–0·04% and healthy leaves 0·8–8·5% Cl. Sauchelli[301] reports that the leaves take on a mottled chlorosis which becomes visible only in transmitted light; these affected areas later appear light green and are depressed. There are no reports of field deficiencies of this element because rain brings far more chlorine than is needed everywhere and plants always contain large excesses.

COPPER

Deficiency symptoms

Van Schreven[305] first demonstrated the symptoms which characterise copper deficiency in sugar beet by growing plants in carefully purified nutrient solution. The symptoms appeared after seedlings had been growing in the solution without copper for three weeks. The heart leaves stayed green but the older leaves became chlorotic in between the veins. The veins remained dark green, contrasting with the light yellow areas. Some older leaves were a blue–green colour. Root development was also retarded by copper deficiency and yields were much decreased, as was the sugar percentage. Another characteristic of copper-deficient plants was that the necrotic areas on old leaves were greyish-brown or greyish-white and the dead leaves looked bleached.

Ulrich and Hills[347] confirmed these symptoms and found they were able to produce symptoms only after thorough purification of salts and water used in nutrient culture, suggesting that sugar beet needs extremely small amounts of copper to sustain growth. Pizer et al.[278] and Hull[187] found no evidence of copper deficiency symptoms in sugar-beet crops in England. This has been confirmed for sugar beet in USA by Ulrich and Hills[347] and Haddock and Stuart,[147] but Pizer et al.[278] did find that some crops responded to treatment with copper even when no plants showed signs of deficiency (see below).

Copper in the plant

In a survey of sugar beet in Western USA, Haddock and Stuart[147] found that the dried leaf blades contained up to 20 ppm. None of the crops showed symptoms of deficiency and the authors concluded that crops were adequately supplied when the blades contained more than 7 ppm.

Sauchelli[301] found that dried tops of plants which were copper deficient contained 6 ppm Cu and treatment with 100 lb/acre copper sulphate increased yield. However, the concentration in the tops only increased to 7 ppm Cu. Pizer et al.[278] found that ten years of cropping (which included two sugar-beet crops) only removed 0·5 lb/acre Cu. Sugar-beet roots in Britain contain 0·85–1·09 ppm Cu.

Copper in soil

Pizer et al.[278] investigated copper deficiency in crops in Eastern England and suggested deficiency was likely when the soil contained the following concentrations:

	Cu ppm
Organic soils	0–4
Loamy sands	0–1·3

They extracted the copper from air-dried soil with the ammonium salt of ethylene diamine tetra acetic acid (EDTA) at pH 4. Sugar beet gave a large response to copper-sulphate applied to an organic soil when it contained 3 ppm. The treatment was effective in increasing soil copper for more than ten years.

FIELD EXPERIMENTS WITH COPPER

Lachowski[211] tested 27 lb/acre copper sulphate applied for sugar beet in several regions of Poland. It had little effect on yield on loams and sandy clay podsols. On some fields it gave a slight increase, probably because it controlled *Cercospora beticola*.

Pizer et al.[278] reported worthwhile increases in yield from copper applications on some soils in Eastern England but they never saw any deficiency symptoms; both sprays and soil applications were tested. The soil treatment (33 lb/acre copper sulphate, 8·4 lb/acre Cu,

applied three years before) increased root yield. A foliar application of $4\frac{1}{2}$ lb/acre oxychloride (2 lb/acre Cu) checked the growth of
tops for a short period but increased the sugar yield (cwt/acre):

No copper	Seedbed application	Foliar application	Seedbed and foliar application
36	41	39	43

Response to sprays was less the later they were applied and the best
time of application was when the plants had 8–10 leaves (1 July).

IRON

Deficiency symptoms

IN THE FIELD

Van Schreven[305] first reported that iron deficiency occurred on
some alkaline soils. Wallace[353] confirmed the finding and Hull[187]
described the symptoms. Deficiency symptoms occur sporadically,
usually in May or June on sandy calcareous soils. In some years
numerous fields in East Anglia have plants with symptoms in May
or June, particularly on sandy calcareous soils. Characteristically
the symptoms disappear quickly when weather conditions change.
In Nebraska, USA, Peterson et al.[275] found that the iron content
of sugar beet was adequate although in a survey a wide range of
total iron concentrations was observed in sugar-beet leaves. The
amounts of soil iron extracted by 0.1 N HCl were not related to
plant uptake in the survey.

IN THE GREENHOUSE

Hewitt[176] found that sugar-beet plants in sand culture given iron
as citrate or magnetite grew normally. Without iron, plants developed
interveinal mottling in young leaves followed by acute chlorosis and
interveinal necrosis. Nagarajah and Ulrich[252] found that plants in
culture solution without iron developed chlorosis first on the young
leaves in the heart of the plant. Severely affected leaves became
completely bleached and developed necrotic spots. On recovery,
such leaves formed a network of prominent green veins which often
characterises iron deficiency in the field.

When plants are grown in water culture, iron deficiency symptoms
may be caused by excessive concentrations of manganese salts.

Hewitt[175] confirmed this effect with sugar-beet plants but found that metals other than manganese also caused iron deficiency. He showed that the toxic effects of excess manganese could be readily distinguished from true iron deficiency by the nature of the symptoms.

Response to iron

In the greenhouse
Nagarajah and Ulrich[252] have reported one of the few investigations of the iron nutrition of sugar beet. The plants were grown in culture solution with 11 different amounts of iron (from 0·20 to 200 mg Fe/20 1). Plants were harvested when those in the smallest five of the 11 iron treatments showed iron deficiency symptoms.

Figure 19 shows that leaf yield was decreased when the concentration of iron in the dry matter of the laminae was less than 55 ppm.

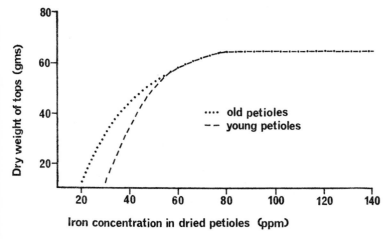

FIG. 19.   Concentration of iron in dried petioles and the weight of tops in culture solution.[252]

The stage of maturity of the leaf sampled had little effect. Laminae ranging in symptoms from severe yellowing and necrosis to a light green contained from 20 to 50 ppm Fe in the dry matter. Normal laminae contained from about 60 to 150 ppm.

In the field
The influence of iron sulphate ($FeSO_4 . 7H_2O$) on the yield of sugar beet in Poland was investigated in 82 field experiments by Lachowski

and Wesolowski.[212] At 36 lb/acre iron sulphate increased the yield of roots by 10% and the tops by 5%. Yield was only increased on podsolic soils containing less than 0·45% Fe in the plough layer. The dressing also increased sugar percentage by about 0·3% on deficient soils but had no effect on emergence, bolting or virus yellows incidence.

MANGANESE

Deficiency symptoms

Leaves of plants deficient in manganese show symptoms which are easily recognised. They have a characteristic upright posture due to the petioles growing almost vertically out of the crown and the laminae rolling inwards. Leaves of affected plants are also chlorotic in between the veins, these areas being pale yellow or almost white. As the leaves age, small angular and sunken spots appear which gave rise to the term 'Speckled Yellows' first being used to describe what is now known to be manganese deficiency. These small necrotic speckles are brownish pink and translucent. Later the necrotic

TABLE 58

AREA OF SUGAR BEET IN GREAT BRITAIN AFFECTED BY TRACE ELEMENT DEFICIENCY SYMPTOMS
(from British Sugar Corporation fieldmen's records)

| | Means, 1961–65 | | | |
	June	July	August	September
	(1 000 acres)			
Boron	0	2	7	4
Manganese	19	7	4	1
	(1 000 ha)			
Boron	0	1	3	2
Manganese	8	3	2	0

| | Means, 1961–70 Percentage of plants affected | | | |
	1–20	21–60	61–100	Total
	(1 000 acres)			
Boron	8	1	0	9
Manganese	25	8	4	37
	(1 000 ha)			
Boron	3	0	0	3
Manganese	10	3	2	15

areas may fall out, leaving leaves with many small holes and when the plants have recovered these are seen against a background of green healthy-looking leaf.

In Great Britain symptoms appear most commonly on plants from May onwards and disappear in August, although they may appear and disappear at any time. The severity of the symptoms fluctuates during this period but, on average, declines from June to September (Table 58). During the period 1961–65 the average acreage affected by manganese deficiency, as recorded by the fieldmen of the British Sugar Corporation, decreased from 19 000 acres in June to 7 000 in July and to 1 000 in September. Table 58 also shows that the average annual total acreage affected in Great Britain was 37 000 acres. Of this 25 000 acres was only slightly affected (1–20% of plants with symptoms), 8 000 was moderately affected (21–60%) and only 4 000 severely affected (61–100%).

OCCURRENCE

Manganese deficiency normally occurs only when the soil pH is greater than 6·5 and then only when there is a large proportion of organic matter present. Severe deficiency occurs regularly only on peaty soils of the fens in Great Britain where the organic matter content of the soil is above about 20%, although severe symptoms are reported sporadically from other regions. Waterlogging of soil excludes oxygen, so increasing the availability of tri- and tetra-valent manganese by aiding its reduction to the divalent form. Oxidising bacteria and organic matter reverse this process. Slight deficiency symptoms appear occasionally in most sugar-beet growing areas, particularly after heavy liming or in ploughed-down pasture. Compared with other crops, sugar beet is particularly sensitive to shortage of manganese.

PLANT MANGANESE AND DEFICIENCY SYMPTOMS

Table 59 summarises concentrations of manganese found in sugar-beet plants (mainly from Great Britain) with and without deficiency symptoms. Concentrations of manganese in sugar beet vary more than for any other trace element. Leaves with symptoms usually contain only 10–20 ppm Mn, whereas leaves from healthy plants of the same age contain 50–200 ppm.

Hale et al.[150] followed the changes in plant manganese after application of manganese sulphate at 0, 5, 10, 20 and 40 lb/acre (Table 60). There was a decrease in symptoms with increasing rate of application of manganese, accompanied by an increase in the manganese concentration in the leaves. Leaves were sampled on three dates and it was noticeable that the manganese concentration

TABLE 59

MANGANESE CONCENTRATION IN PLANTS WITH AND WITHOUT
DEFICIENCY SYMPTOMS

Plant part	With symptoms	Without symptoms	References
	(ppm Mn in dry matter)		
Leaf	14	53	Morley Davies[250]
Leaf	17	46	Wallace[353]
Laminae	36	50–178	Larsen[216]
Petioles	15	20–46	
Leaf	12	110	Hale et al.[150]
Petioles	10	15–25	Vömel and Ulrich[351A]
Laminae	4–20	25–360	Ulrich and Hills[347]
Leaf	12–17	46–110	Range
Leaf	14	70	Means

diminished rapidly with time. The authors suggest that this was
because the manganese was immobilised in the sprayed leaves and
steadily lost by death of those leaves, a suggestion which has since
been confirmed (see below).

There are few reports of the total uptake of manganese by sugar
beet. On average, a 15 ton/acre crop of sugar-beet roots removes
about 0·75 lb/acre Mn; tops remove about 1 lb/acre.

TABLE 60

EFFECT OF MANGANESE SPRAY ON MANGANESE
CONCENTRATION IN SUGAR-BEET LEAVES
(after Hale et al.[150])

Manganese sulphate applied on 5 July		Mn in dried leaves (ppm)		
(lb/acre)	(kg/ha)	July	August	September
0		20	10	16
5	6	105	21	19
10	11	172	29	20
20	22	350	55	20
40	45	560	79	26

Movement of manganese in the plant

Henkens and Jongman[169] used radio-active manganese (Mn^{54}) to
investigate translocation of the element in sugar-beet plants. In
fact, their experiments demonstrate very well that it does not move

from leaf to leaf. Figure 20 shows the concentration of manganese in sugar-beet leaves from plants where one leaf was treated with manganese solution compared with leaves from untreated plants. After 20 days only the treated leaf contained the applied manganese. The manganese did, however, move down into the roots.

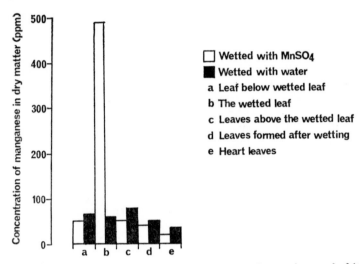

FIG. 20.   Manganese concentration of leaves 20 days after wetting one leaf.[169]

Field experiments were made which confirmed these findings; newly-formed leaves did not benefit from spray given earlier, with the result that deficiency symptoms reappeared. One spray scarcely increased yield, whereas two sprays gave substantial increases of over 10%.

Response by sugar beet to manganese applications

In Great Britain
Hull[190] made the first comprehensive study in Great Britain in 1935/40; i.e. six field experiments on an organic fen soil tested seedbed and top-dressings of solid and foliar-applied solutions of manganese sulphate ($MnSO_4 . 1H_2O$—33% Mn). On average, seedbed manganese increased root yield by 0·5 ton/acre. In three experiments there was a response of about 1 ton/acre and little response in the others. On average, 50 and 150 lb/acre of manganese sulphate were equally effective, but the large amount was justified where the crop was

deficient. Sugar percentage was increased slightly by the manganese treatments and the average increase in sugar yield was 2·5 cwt/acre.

Where solid manganese sulphate was applied as a top-dressing in June it was marginally less effective than the equivalent seedbed dressing. Sprays of manganese sulphate solution applied to the leaves also increased yield less than the seedbed dressing. Where symptoms were severe (Mn concentration in leaves 14 ppm) treatment increased yield by 7 cwt/acre sugar.

IN HOLLAND

Henkens and Smilde[171] investigated the value of manganese sulphate compared with the element in the form of silicate 'frits'. Both increased reducible soil manganese and the effect lasted for 1½–2 years. There was no clear relationship between reducible (considered by some as 'plant-available') soil manganese and the occurrence of deficiency. However, the materials increased yield, sugar production (Table 61) and leaf manganese, whilst decreasing the incidence of

TABLE 61

RESPONSE BY SUGAR BEET TO SOIL APPLICATIONS OF
MANGANESE SULPHATE IN HOLLAND
(after Henkens and Smilde[171])

Manganese sulphate	Root yield	Tops yield	Sugar yield
(lb/acre)	(ton/acre)		(cwt/acre)
0	23	12	66
89	25	13	74
178	25	14	73
357	26	15	77
(kg/ha)		(t/ha)	
0	57	31	8·3
100	64	32	9·3
200	63	35	9·2
400	65	37	9·6

deficiency symptoms. When the two materials were applied to the soil they were more effective than a foliar spray of manganese sulphate. With increasing quantities of fertiliser (90–360 lb/acre manganese sulphate or 76–640 lb/acre frit) there was no significant increase in yield and sugar production (Table 61), whereas leaf manganese and incidence of manganese deficiency only responded to the large dressings. None of the treatments controlled the symptoms throughout the whole season. The percentage of manganese-deficient plants was related to both leaf and reducible soil manganese (Fig. 21) but not to yield.

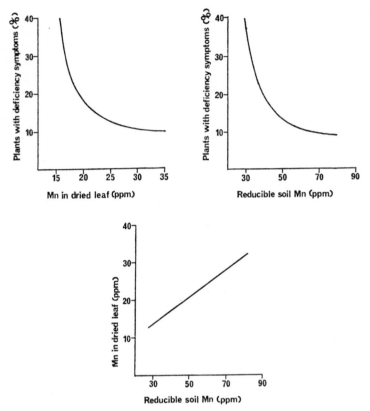

FIG. 21. Inter-relationships between plant and soil manganese and the percentage of plants with deficiency symptoms.[171]

Effect of other fertilisers on manganese deficiency

Hale *et al.*[150] found that sulphate of ammonia, potash and salt decreased the symptoms and increased the manganese concentration in the leaves. Phosphate also increased the concentration of leaf manganese as below:

Effect of	Leaf Mn (ppm)
N	+29
K	+9
Na	+11
P	+20

Peterson *et al.*[275] in Nebraska thought that phosphorus did not increase manganese uptake, and others have found that manganese uptake by plants is *decreased* by phosphate application on many neutral and acid soils. On such soils, plants still have sufficient manganese for healthy growth because its availability is usually only restricted above pH 6·5. Where manganese was deficient, Larsen[216] found that phosphate *increased* the availability of manganese. Triple superphosphate given for sugar beet increased the manganese concentration of the leaves from 36 to 178 on one field and from 60 to 106 ppm on another field. In the first case the phosphate cured manganese deficiency (there was none in the second case) and increased yield of roots and tops and the manganese and phosphate concentration of the leaves. However, spraying with manganese sulphate solution did not increase the yield or the manganese concentration of the leaves.

Several theories are put forward to explain this beneficial effect of phosphorus fertiliser on manganese-deficient sugar beet: (1) a chemical mobilisation of soil manganese by the phosphorus; (2) a microbiological mobilisation; (3) improved root development leading to exploitation of a greater soil volume; (4) increased ability of roots to absorb manganese brought about by phosphorus uptake; (5) improved translocation of manganese inside the plants. Larsen suggests that (1) is the most likely factor which was operating. Acid phosphates may simply dissolve manganese thus increasing its availability, whereas neutral or alkaline salts precipitate it.

MOLYBDENUM

There have been few investigations of the molybdenum nutrition of sugar beet. Ulrich and Hills[347] described the symptoms of deficiency and found that healthy leaves contained 0·2–20·0 ppm Mo in dry matter, whereas deficient leaves contained 0·01–0·15 ppm. Henkens and Smilde[170] found increased yields in pot experiments from molybdenum given to sugar-beet plants grown in molybdenum-deficient soil (previous sugar-beet plants grown in the same soil showed distinct symptoms of molybdenum deficiency). Sodium molybdate ($Na_2MoO_4 \cdot 2H_2O$—39·6% Mo) was tested and compared with several glassy 'frits' (2 or 3% Mo). The amounts tested were between 0·05 and 0·70 lb/acre Mo.

Dry matter production was increased considerably by molybdenum and there was a negative relationship between dry matter yield and severity of molybdenum deficiency. Sodium molybdate gave the largest yield and plants receiving the equivalent of 0·54 or 0·70 lb/acre Mo had no symptoms of the deficiency. The frits varied in ability to correct deficiency but had a greater residual effect than the molybdate, as shown by further cropping.

Nowicki[257] investigated the effect of different quantities of molybdenum fertiliser on the yield, health and processing quality of sugar beet grown on acid, neutral and alkaline soils in Poland. It had no effect on yield but showed some tendency to decrease the incidence of disease and increase the quality of the roots.

RUBIDIUM

Potassium is the only univalent cation generally recognised as being indispensable for growth of all plants. However, it has been shown in Chapter 4 that sodium (whether or not it is indispensable for growth) is of great economic importance in the sugar-beet crop. Rubidium, another univalent cation, is very similar to potassium. El-Sheikh et al.[118] investigated the effects of rubidium on sugar beet in water culture and its interactions with sodium and potassium. It increased the growth of plants significantly when supplied in small doses when the plants were deficient or adequately supplied with potassium. Although the authors inferred that rubidium was an essential nutrient for maximum yield, large concentrations were toxic, especially to the growth of roots.

ZINC

There is no evidence of zinc deficiency in sugar-beet crops in Great Britain but it has been reported in some areas of the world. Maize is an indicator crop, showing characteristic symptoms when the supply of zinc from the soil is short. Sugar beet on the same fields occasionally shows symptoms of deficiency, but the crop is less sensitive than maize and field beans.

Boawn and Viets[27] described zinc deficiency on certain soils in Washington State. The soils were very alkaline (pH 9·5) and sugar beet responded to treatment with zinc sulphate broadcast at 16 lb/acre Zn prior to seedbed preparation in spring. Where no zinc was given, plants were smaller, the leaf blades were decreased in size with yellow margins, and the tips were particularly yellow. Zinc sulphate cured the symptoms. The zinc concentration of deficient leaves was 8 ppm and 13 ppm in healthy leaves.

In a later paper Boawn et al.[28] found that 16 lb/acre Zn, applied two years before sugar beet, increased the zinc concentration of leaf blades from 20 to 30 ppm, of tops from 12 to 22 ppm, and of roots from 8 to 12 ppm. These increases in the concentration of zinc in the plant did not increase yield, but the total zinc contained in a crop yielding 30 tons/acre varied from 0·18 to 0·27 lb/acre depending on the amount of zinc given. A survey carried out by Peterson *et al.*[275] showed that the soils in the major sugar-beet producing area of Nebraska could supply an adequate amount of zinc for maximum yield. They also showed that the concentration of zinc in the plant (10–90 ppm in petioles) was significantly related to the amount of zinc extracted from the soil by 0·1 N HCl (2–7 ppm).

Lachowski[210] tested the effect of zinc sulphate on sugar beet in 82 experiments in Poland. In the north of the country on sandy clay podsols, 4·5 lb/acre zinc sulphate ($ZnSO_4 \cdot 7H_2O$) gave the largest yield and sugar percentage but on loamy and black soils in other regions 9–18 lb/acre was needed. On potassium-rich soils, zinc had little effect on yield.

Rosell and Ulrich[290A] determined the critical concentration of zinc in sugar-beet plants in nutrient solutions; this was 8 to 10 ppm for dried mature leaf blades. They described the deficiency symptoms in nutrient culture. 'Symptoms first appeared on young leaves which grew erect and turned light green. As the chlorosis became more pronounced the interveinal areas of the upper surface were pitted with white spots. The spotting gradually became worse and turned brown. Shortly afterwards the necrotic interveinal areas deteriorated rapidly and the entire leaf collapsed.' Giving zinc to healthy plants decreased the concentration of iron in the leaves and the zinc-deficient plants contained up to ten times more iron than normal leaves.

Chapter 8

Organic Manures and Green Manuring

Organic manures are applied before growing sugar beet in many countries. In Great Britain, between a fifth and a quarter of the crop is grown with an organic manure, but the area treated is decreasing.[84] Most of the manure used consists of a mixture of partly-rotted straw and animal residues, but poultry droppings, wool-waste (shoddy) and sewage sludge are also used on a limited scale in some areas. Cooke[66] has comprehensively reviewed the value of organic manures in British agriculture and their effects on yield of many crops.

The farming community seems to consider organic manures mainly as soil conditioners but little account is taken of their nutrient content. For example, Boyd[33] found that the amount of mineral fertiliser used on sugar-beet fields was the same with and without farmyard manure (Table 62). Many one-year experiments in Great Britain and

TABLE 62

AVERAGE QUANTITY OF FERTILISER APPLIED TO FIELDS
WITHOUT AND WITH FARMYARD MANURE ON 51 FARMS
(after Boyd[33])

| | Fertiliser used for sugar beet | | | | | |
| | N | P_2O_5 | K_2O | N | P | K |
		(cwt/acre)			(kg/ha)	
Without farmyard manure	0·92	0·84	1·17	116	46	122
With farmyard manure	0·92	0·83	1·12	116	45	117

abroad have investigated the mineral fertiliser equivalent of farmyard manure; also the optimal dressing of mineral fertilisers for sugar beet when farmyard manure is given. The value of organic manure as a soil-improver—both physically and chemically—has also been tested in a few annual and long-term rotation experiments with sugar beet.

TABLE 63

FERTILISER EQUIVALENTS OF FARMYARD MANURE FOR SUGAR BEET

Farmyard manure applied		Fertiliser equivalent						References
(ton/acre)	(t/ha)	N	P_2O_5 (cwt/acre)	K_2O	N	P (kg/ha)	K	
12	30	0·40	0·40	1·60	50	22	167	Adams[8]
12	30	0·30	—	>0·95	38	—	>99	Draycott[84]
12	30	<0·30	—	—	<38	—	—	
10	25	—	0·40	0·60	—	22	63	Crowther and Yates[71]
10	25	>0·30	>0·42	>0·75	>38	>23	>78	Patterson and Watson[266]
?	?	0·16–0·48	—	—	20–60	—	—	Jorritsma (Holland)[199]
Mean 11	28	0·32	0·41	0·98	40	23	78	

ORGANIC MANURES

Fertiliser equivalent of farmyard manure

Experiments testing nitrogen, phosphorus, potassium and, more recently, sodium, with and without farmyard manure, allow the fertiliser equivalent of farmyard manure to be estimated for the sugar-beet crop. Adams[8] described over 40 experiments testing a factorial arrangement of nitrogen, phosphorus and potassium all with and without 12 tons/acre farmyard manure applied before ploughing; 0·6 cwt/acre N decreased sugar percentage by 0·4% and farmyard manure by 0·2%. On this basis, farmyard manure provided about 0·30 cwt/acre N. However, in yield of tops the farmyard manure was equivalent to 0·45 cwt/acre N and in terms of juice purity equivalent to 0·60 cwt/acre N. With farmyard manure, sugar yield was not increased by more than 0·8 cwt/acre K_2O, so the farmyard manure provided up to 1·60 cwt/acre K_2O. Farmyard manure could, however, have given this response partly by providing sodium, which greatly decreases response to potassium (page 74). Responses to phosphorus were small both with and without farmyard manure, but the farmyard manure seemed to be equivalent to about 0·4 cwt/acre P_2O_5 (Table 63).

More recent experiments[84] made on commercial farms tested nitrogen, agricultural salt (sodium chloride) and 12 tons/acre farmyard manure. A basal dressing of phosphorus and potassium was

TABLE 64

DRY MATTER PERCENTAGE, NUTRIENT CONCENTRATION AND QUANTITY OF NUTRIENT IN FARMYARD MANURE APPLIED FOR SUGAR BEET: MEANS OF 17 SAMPLES

(after Draycott[84])

Dry matter (%)	Nutrient concentration in dry matter (%)				
	N	P	K	Na	Mg
26·0	2·46	0·72	2·45	0·44	0·68

Quantity of nutrient in 12 ton/acre dressing (cwt/acre)				
N	P_2O_5	K_2O	Na_2O	MgO
1·49	1·01	1·64	0·35	0·55

Quantity of nutrient in 30 t/ha dressing (kg/ha)				
N	P	K	Na	Mg
187	55	171	16	41

given to all the plots. In terms of sugar yield, farmyard manure was 'worth' 0·30 cwt/acre N; it also prevented response to agricultural salt due to the potassium and sodium it contained (Table 63). Samples of the farmyard manure used in the experiments were analysed for total concentrations of major nutrients and the average results are shown in Table 64. In terms of sugar yield, only about a fifth of the nitrogen was available to the crop but most of the potassium and sodium.

Optimal fertiliser dressings where farmyard manure is used

Some of the experiments described above were made to determine optimal fertiliser dressings for sugar beet on fields given farmyard manure. Such information is important in Great Britain for each year about 100 000 acres of sugar beet are grown with farmyard manure. Crowther and Yates[71] made the first thorough investigation and in a survey of all the British and some other European data they found that the optimal dressings with farmyard manure were 0·70 cwt/acre N, 0·60 cwt/acre P_2O_5 and 0·30 cwt/acre K_2O. Without farmyard manure the same amount of nitrogen was considered optimal but an additional 0·30 cwt/acre P_2O_5 and 0·35 cwt/acre K_2O was needed.

Patterson and Watson[266] tested 0, 5, 10 and 20 tons/acre farmyard manure for sugar beet and potatoes, with and without nitrogen, phosphorus, potassium and sodium in eight experiments at Rothamsted and Woburn. Table 65 shows that the responses to fertilisers were all decreased by farmyard manure and that decreases were greatest when responses were greatest. There was a particularly large response to nitrogen at Woburn, and a large decrease in the response when farmyard manure was also given. They also found that farmyard manure was more effective relative to fertilisers for sugar beet than for potatoes, but no explanation of the effect was given.

Adams[8] found that the optimal dressing on plots given 12 tons/acre farmyard manure was 0·6 cwt/acre N, no P_2O_5, and 0·8 cwt/acre K_2O, whereas on plots without farmyard manure the best dressing was 1·0 cwt/acre N, 0·5 cwt/acre P_2O_5 and 1·6 cwt/acre K_2O. Sodium affects both nitrogen and potassium requirement of sugar beet but it was not included in Adams' experiments and further experiments[84] tested sodium in addition to the other elements on fields given a basal dressing of 12 tons/acre farmyard manure. In these experiments the optimal dressing of fertiliser was 0·6 cwt/acre N, 0·3 cwt/acre P_2O_5 and 0·5 cwt/acre K_2O; on most fields sodium only gave a small increase in yield, presumably due to the sodium and potassium in the farmyard manure. However, it largely replaced the need for potassium, so where farmyard manure and agricultural salt are given potassium is rarely needed.

Boyd[34] thought that farmyard manure acted independently of nitrogen, the optimal dressing of nitrogen (1·0 cwt/acre) for sugar yield being unchanged by farmyard manure (Fig. 22a). More recent experiments[84,38] indicate quite a different relationship; it is no longer thought that the substantial decrease in sugar yields from large nitrogen dressings are likely (Fig. 22b).

TABLE 65

EFFECT OF FARMYARD MANURE ON RESPONSE TO FERTILISERS BY SUGAR BEET AT ROTHAMSTED AND WOBURN: MEANS OF FOUR EXPERIMENTS AT EACH STATION
(after Patterson and Watson[266])

| Fertilisers applied | Mean effect of fertilisers without farmyard manure | | Mean effect of fertilisers with farmyard manure | |
	Rothamsted	Woburn	Rothamsted	Woburn
		Sugar yield response		
(cwt/acre)		(cwt/acre)		
0·9 N	−0·8	12·1	−0·7	−4·1
1·75 P_2O_5	10·9	−0·1	−2·6	−1·2
1·5 K_2O	2·6	0·8	−0·8	−0·3
5 NaCl	5·4	4·1	−1·5	−2·2
(kg/ha)		(t/ha)		
113 N	−0·10	1·52	−0·09	−0·52
96 P	1·37	−0·01	−0·33	−0·15
156 K	0·33	0·10	−0·10	−0·04
250 Na	0·68	0·52	−0·19	−0·28

Response to nitrogen in the presence and absence of farmyard manure depends on the basal dressing of other nutrients, as first noticed by Boyd.[33] When little or no phosphorus or potassium is applied, response to nitrogen alone is limited—it is easy to draw the false conclusion that farmyard manure does not decrease response to nitrogen (in fact, in some cases it increases response). This appears to be the reason for Crowther and Yates'[71] report that the same amount of nitrogen should be used whether or not farmyard manure is given. The most recent experiments[84] lead to the conclusion that when crops are given an adequate supply of other nutrients the nitrogen dressing should be decreased by 0·30 cwt/acre. The same principle applies to other nutrients.

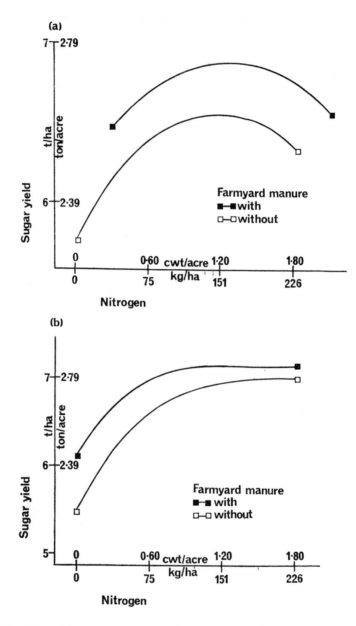

FIG. 22. Effect of farmyard manure on response to nitrogen: (a) ref. **34**; (b) ref. 84.

Effects of other organic manures on sugar beet

SEWAGE SLUDGE

Sewage sludges contain much organic matter but, like farmyard manure, their main agricultural value is in the nutrients they supply, not the organic matter. Cooke[66] has reviewed responses by many crops to sewage sludge and compared them with composts and farmyard manure.

Numerous experiments have tested sludges for sugar beet and most have included comparisons with farmyard manure and fertilisers. Bunting[46] in three experiments at Saxmundham (on a chalky boulder-clay soil) found that sewage sludge increased yield of sugar-beet roots from 15 to 16 tons/acre whereas farmyard manure increased it to 18 tons/acre. The sludge was less effective than farmyard manure, probably because it contained much less potassium (to which sugar beet gives a large response) than farmyard manure but more phosphorus (to which sugar beet gives relatively little response on most soils). Table 66 shows analyses of some of the materials used by Bunting, which are typical.

TABLE 66

AVERAGE CHEMICAL COMPOSITION OF SOME ORGANIC
MANURES
(after Bunting[46])

	Dry matter (%)	Organic matter	As percent dry matter			
			Total N	Inorganic N	P	K
Raw sludge	40	51	2·4	0·13	0·56	0·25
Digested sludge	52	44	2·6	0·12	0·96	0·33
Dried sludge	80	39	2·4	0·33	1·14	0·33
Straw/sludge compost	30	53	2·1	0·09	0·66	0·58
Farmyard manure	26	64	2·2	0·22	0·70	1·66

Garner[127] compared digested sewage sludge, mostly of domestic origin, with farmyard manure for four years. Fertilisers were also tested and the cumulative effect of all the treatments measured in a final crop of sugar beet. Farmyard manure was tested at 8 and 16 tons/acre and sludge at 5 and 10 tons dry matter per acre (about 10 and 20 tons fresh sludge). Table 67 shows the effect of the organic manures and the fertilisers. Both organic manures increased yield of roots and sludge gave about twice the yield of tops as farmyard manure. The fertiliser dressings (0·6 cwt/acre of N, P_2O_5 and K_2O) were too small for maximum yield but Table 68 shows the interactions between organic manures and fertilisers, confirming that

TABLE 67

RESPONSE BY SUGAR BEET TO FERTILISERS IN THE ABSENCE OF
ORGANIC MANURE AND TO SINGLE AND DOUBLE DRESSINGS OF
FARMYARD MANURE AND SEWAGE SLUDGE
(after Garner[127])

| | Roots | Tops | Roots | Tops |
	(ton/acre)		(t/ha)	
Yield with no organic manure	11·6	10·0	29·2	25·1
Increase from:				
Ammonium sulphate	0·7	3·3	1·8	8·3
Superphosphate	1·0	0·6	2·5	1·0
Potassium chloride	0·1	−0·2	0·3	−0·5
Single farmyard manure	1·1	1·4	2·8	3·5
Double farmyard manure	1·9	2·5	4·8	6·3
Mean effect of farmyard manure	1·5	1·9	3·8	4·8
Single sludge	1·2	3·0	3·0	7·5
Double sludge	2·2	5·0	5·5	12·6
Mean effect of sludge	1·7	4·0	4·3	10·0

both manures were acting mainly as a source of these three major
nutrients—farmyard manure largely as a source of available potassium and sludge as a source of nitrogen. In the final year the cumulative effect of four applications of all treatments was measured. The
increases in yield of sugar beet were 1·4 tons/acre roots from 48

TABLE 68

MEAN INCREASES IN YIELD FROM FARMYARD MANURE AND
SEWAGE SLUDGE IN PRESENCE AND ABSENCE OF FERTILISERS
(after Garner[127])

| | Increase in root yield above basal fertiliser from: | | | |
| | Farmyard manure | Sludge | Farmyard manure | Sludge |
	(ton/acre)		(t/ha)	
Nitrogen:				
Absent	1·1	2·4	2·8	6·0
Present	1·9	1·0	4·8	2·5
Difference	0·8	−1·4	2·0	−3·5
Phosphorus:				
Absent	2·0	2·1	5·0	5·3
Present	1·0	1·3	2·5	3·3
Difference	−0·9	−0·8	−2·3	−2·0
Potassium:				
Absent	1·1	1·8	2·8	4·5
Present	−0·1	1·6	−0·3	4·0
Difference	−1·2	−0·2	−3·0	−0·5

tons/acre farmyard manure and 0·7 tons/acre from 60 tons/acre sewage sludge.

Vaskhnil and Manorik[350] made experiments in Russia with organic wastes for sugar beet. Farmyard manure and other organic materials were improved by composting with fertiliser. Additions of trace elements, azotobacteria and phosphobacteria during the process also improved the fertiliser value of the compost. The best period for composting was 6–7 months, after which the available nitrogen in the compost diminished.

STRAW AND STRAW COMPOSTS

Patterson[265] reported a three-course rotation experiment at Rothamsted designed to investigate the benefits, if any, from ploughed-in or composted wheat straw. Potatoes, barley and sugar beet were grown, and the yields of sugar from the beet are shown in Table 69.

TABLE 69

EFFECT OF FERTILISER AND STRAW PLOUGHED IN AND COMPOSTED ON YIELD OF SUGAR: MEAN OVER 18 YEARS AT ROTHAMSTED
(after Patterson[265])

	Applied to: test crop	Applied to: previous crop	Applied to: test crop	Applied to: previous crop
	(cwt/acre)		(t/ha)	
Fertiliser only (F)	43·3	37·3	5·43	4·68
Straw ploughed in, fertiliser in spring (Ss)	41·0	37·4	5·15	4·69
Straw ploughed in, half fertiliser in autumn, half fertiliser in spring (Sd)	40·9	38·6	5·13	4·62
Straw composted with fertiliser and ploughed in (C)	36·9	36·1	4·63	4·53

The four treatments were: F, fertilisers only, applied in spring (0·4 cwt/acre N, 0·4 cwt/acre P_2O_5 and 0·5 cwt/acre K_2O); Ss, 53·3 cwt/acre straw ploughed in plus F in spring; Sd, as Ss but with half the fertiliser applied in autumn; C, as Ss but straw composted with the fertiliser. The treatments were applied to four plots in odd years and a further four plots in even years. In this way both direct and residual effects could be measured.

When straw is ploughed in, nitrogen additional to that contained in the straw itself is required for aerobic decomposition. Much of this nitrogen is converted into unavailable or slowly available forms. Patterson explained his results largely on this basis, for straw ploughed in decreased yields in the first year and where no fertiliser was given in spring (as when the straw was composted) yield was reduced even further. After 18 years of the original treatments the scheme was changed for a final six years and additional nitrogen and potassium tested. Only 0·2 cwt/acre N applied in spring was needed to compensate for the losses in yield due to ploughing straw into the land. Nitrogen deficiencies were less serious for straw ploughed in directly than for compost. This may have been due to the straw using soil nitrogen which would otherwise have been washed away by winter rain. The treatments had no effects on crop yields through improvements to the soil structure or increased organic matter.

Experiments similar to the ones at Rothamsted were begun in 1936 at the Norfolk Agricultural Station.[162,286] Crops were grown for 20 years in a rotation somewhat like the traditional Norfolk Four Course—sugar beet, barley, one-year ley and wheat. The three treatments were farmyard manure (11 tons/acre/rotation), straw (55 cwt/acre) and no organic manures. When straw was given, 0·15 cwt/acre N was also given. On average, farmyard manure increased sugar-beet root yield by 3 tons/acre and straw by 0·4 tons/acre. These improvements in yield seem to have been largely a result of the poor fertility maintained throughout the experiments, for only 0·70 cwt/acre N, 0·54 cwt/acre P_2O_5 and 0·36 cwt/acre K_2O were given per rotation. There was little evidence that responses to the organics increased the longer the experiments continued. Soil analyses indicated that both phosphorus and potassium in farmyard manure plots had increased but that there was little change in plots with straw, although the organic matter content increased by 0·1 % over the 20-year period. Thus these experiments confirm the findings of the Rothamsted experiments—where small amounts of fertilisers are given, it is the nutrients supplied by ploughed-in straw or farmyard manure which increase yields of sugar beet, *not* the organic matter.

ORGANIC WASTES

Domestic waste in various forms was tested by Garner[127] for sugar beet and other crops at Rothamsted and elsewhere, with farmyard manure as a standard. Analyses of some of the materials tested are given in Table 70. The organic manures were used at 8 and 16 tons/acre with and without nitrogen and potassium. The experiments did not test phosphorus, and differences due to this element were prevented by basal dressings of superphosphate given to all plots.

Pulverised refuses increased sugar-beet yields by only 22 % of the increase from plots with farmyard manure. Screened dust gave variable response when used for sugar beet—it was usually less

TABLE 70

ANALYSES OF TOWN REFUSES APPLIED FOR SUGAR BEET.
PERCENTAGE IN DRY MATTER
(after Garner[127])

	N	Total P	K	Readily soluble K
Screened dust	0·68	0·25	0·27	0·15
Pulverised refuse	0·83	0·25	0·31	0·17
Fermented refuse	0·88	0·20	0·28	0·17
Composted refuse	0·88	0·36	0·35	0·26

useful than farmyard manure or refuse and in some experiments it was harmful. Most organic manures also had residual effects with sugar beet that approached their first-year effects in size. Screened dust, which in the year of application compared unfavourably with the other manures, improved with time and was at least as good as the various pulverised refuses in the final years.

Effects of organic manure other than as a supply of major nutrients

There are numerous reports of experiments where organic manures have given larger increases in yield than equivalent dressings of fertiliser. Frequent, large applications of organic manures cause profound changes in the soil physical conditions and some of these have been measured in long-term experiments considered in the next chapter, but when the beneficial 'extra' effects occur in annual experiments they are not so easily explained.

Both Adams[8] and Draycott[84] in annual experiments testing farmyard manure on 80 farms found that yields of sugar beet from a few fields were greater with farmyard manure and fertiliser than with any amount of fertiliser alone. They concluded that occasionally sugar beet benefits from farmyard manure due to some factor other than nitrogen, phosphorus, potassium and sodium. However, the fertilisers were applied on the surface and the farmyard manure ploughed down, so at least part of the effect may have been due to the farmyard manure supplying nutrients to the crop when those in the fertiliser were less readily available. On some fields the extra effect may have been due to improved supply of magnesium, to trace elements (*e.g.* boron), or to complex chemical effects (*e.g.* plant hormones), all of which topics have been reviewed by Whitehead.[368]

In annual experiments with potatoes, Holliday *et al.*[181] found that responses to farmyard manure were greater in dry than wet years—the responses were linearly related to the maximum soil moisture deficit. They suggested that the nutrients from the organic manure were relatively more available in dry years than nutrients in inorganic fertilisers. No similar study has been made with the sugar-beet crop, but this seems a plausible explanation of some of the responses observed in some annual experiments.

Besides chemical effects, organic manures can act through improvements in physical conditions of the soil. Hoyt[184] measured the effect of farmyard manure and other materials (*see* page 171) on the growth of sugar beet in a light soil. Sufficient nitrogen, phosphorus, potassium and magnesium was also applied to ensure that a shortage did not limit growth. Farmyard manure greatly increased yields of roots but analysis of the plants indicated that the increases in yield were not attributable to improved nutrition. Hoyt therefore concluded that the increases were due to an improvement in the physical condition of the soil, but the nature of the effects was not investigated.

Wellesbourne workers have made numerous investigations of the improvements in soil moisture supply from organic manures in long-term experiments on horticultural soils. Salter *et al.*[300] determined the available water capacity of a fine sandy loam after red beet grown with farmyard manure, sewage sludge or no organic manure. The organic manures had been applied at 6 tons/acre/annum dry matter to the same plots for seven years. The available water capacity was about 30% greater in the farmyard manure plots than the unmanured plots, and 35% greater in the sewage sludge plots. However, a study of the moisture release characteristics of the soils showed the volume of available water was increased little by the manurial treatments although similar studies on a sandy loam soil[299] showed that volume of water available to crops was considerably increased by farmyard manure.

In Bunting's[46] experiments, crops in ten out of 113 cases gave larger responses to organic manures than expected from their nutrient content. All the ten were on light soils. Bunting discounted the beneficial effects of small increases in water supply after calculating that it only amounted to the evaporation needs of 1·5–4 average summer days.

Various other effects of organic manures on sugar beet and related crops have been described—some beneficial, some harmful. Improvements in plant establishment and survival[362] of mangolds as determined by counting the plants before harvest were correlated with increased yield. Farmyard manure (and dug-in fertilisers) improved germination of sugar beet, increased seedling weights and final yield when compared with broadcast fertiliser on Stackyard Field at Woburn.[358] Similar improvements in establishment of globe beet were also reported at Woburn by Mann and Patterson,[240] who also noticed that the beet were of marketable size earliest on the organic manured plots. Two harmful effects of organic manures on beet have been reported—bolting is usually increased by organic manures[240,362] and fanging of the roots increased.

Another aspect of the accumulation of nutrients in soil from repeated applications of organic manures was demonstrated by Warren and Johnston[358] at Woburn. In the Market Garden[240] and the Ley–Arable[36] experiments, phosphorus and potassium accumulated in the soil to a depth of 24 in due to frequent application of farmyard manure, as below:

| Depth (in) | Readily-soluble P and K (ppm) | | | |
| | Fertiliser plots | | Farmyard manure | |
	P	K	P	K
0–9	310	80	850	350
12–18	120	60	210	330
18–24	90	50	170	320

This subsoil enrichment increased yields of red and sugar beet greatly but such frequent, large applications of farmyard manure are far removed from the commercial practice of sugar-beet growing.

GREEN MANURING

In some countries it is common practice to grow sugar beet after a ploughed-in green crop. Many advantages for such a procedure have been claimed and experiments have been made with a wide variety of crops to investigate the value of green manures. The green crop is usually sown under a cereal or direct-sown in late summer or early autumn after a short season cash crop. It is ploughed down in winter or early spring. As with other organic manures, part of the benefit is from the nutrients the green manure supplies and part from improvements in the physical condition of the soil. The crop grown for ploughing down conserves nutrients (e.g. nitrate) by decreasing leaching losses in humid climates. Some green manure crops extract nutrients from the subsoil, increasing the supply in the surface layers for the following crop.

Effect of green manures on sugar beet

IN GREAT BRITAIN

Experiments at Broom's Barn[88] investigated the value of ploughed-down trefoil on the yield and nitrogen fertiliser requirement of sugar beet. The trefoil was under-sown in a barley crop. After the barley harvest, the trefoil grew on for three or four months, fixing atmospheric nitrogen, before the land was ploughed in winter. Chemical analysis showed that the trefoil contained 0·51 cwt/acre N, 0·16

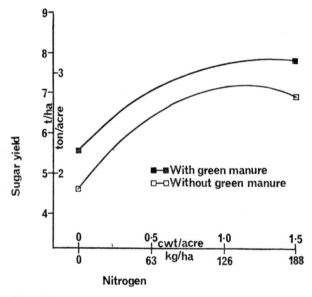

FIG. 23.   Effect of green manure on response to nitrogen at Woburn.[112]

cwt/acre P_2O_5 and 0·52 cwt/acre K_2O. Four amounts of nitrogen fertiliser (0–1·50 cwt/acre) were given in the sugar-beet seedbed and a normal dressing of other major nutrients to all plots. The green manure increased the yield of the sugar beet only when little or no nitrogen was given in the sugar-beet seedbed. Maximum yields from plots with and without green manure were identical provided that the optimum nitrogen dressing was given—this was 0·50 cwt/acre with the green manure and 1·00 cwt/acre without.

Thus at Broom's Barn the green manure acted only as a source of nitrogen when enough of other major nutrients was given and there was no evidence of any other benefits. The quantity of nitrogen in the trefoil at ploughing was a good measure of its nitrogen fertiliser equivalent in terms of sugar-beet yield.

Different results have been reported on the sandy soil at Woburn,[112] where trefoil and ryegrass were both tested. Sugar yield was greater after green manures than with fertiliser alone (Fig. 23). Trefoil increased yield by a similar amount with all the dressings of nitrogen. When no seedbed nitrogen fertiliser was given, trefoil gave the same response as 0·67 cwt/acre N fertiliser, though it contained 0·50 cwt/acre N. Response to ploughed-down ryegrass was governed by the amount of nitrogen previously given to the grass. As with trefoil, the largest yield after ryegrass was greater than the largest yield without green manure.

At Woburn, the green manures appear to act independently of the fertiliser nitrogen. This is an unexpected result and further experiments are in progress to explain it. In the one experiment where roots were counted, the green manures produced a 10–15 % increase in the plant stand. It seems likely that the green manures improved the physical condition of the soil, which improved establishment of the seedlings and increased the growth of the crop.

IN OTHER COUNTRIES

Experiments were made in Poland in the years 1952–57 to explain some aspects of the influence of green manures on sugar beet.[246] The results confirmed earlier experiments,[245] for field peas, spring vetch and field beans ploughed down increased root yield of sugar beet equivalent to two-thirds of the increase from 12 ton/acre of farmyard manure. Leaf yield was greater with green manure than with farmyard manure (which indicates a large supply of nitrogen from the green manure). Sunflower or radish crops were less effective green manures than the legumes. In other experiments, red clover used as green manure gave very good root yields of following sugar beet, and a mixture of sweet clover and yellow clover gave the best leaf yields. They concluded that, on average, a good green manure has a similar fertilising value to an average dressing (8 ton/acre) of farmyard manure. This depends on the quality of the farmyard manure for in Germany, Rauhe and Hesse[285] found that farmyard manure increased sugar-beet yield by 17%, green manure increased it by 36% and the two together by 53%.

Gregg and Harrison[137] reported the effects of ploughing down stubbles of various forage mixtures as green manures before sugar beet in Michigan, USA. The plots had been established for five years before the sugar-beet test crops, which were grown for two years in succession. Sugar beet established best and gave the largest yield after brome grass. Fescues and timothy gave poorest results, while bluegrass and redtop gave intermediate yields. There was a close correlation between the yields of sugar beet and the root habits of the grasses but not with the aggregation or pore space of the soils.

Webb *et al.*[364] made a pot experiment comparing six green manures for sugar beet. Alfalfa, smooth brome grass, Ladino clover, orchard grass, hairy vetch and rye were grown in sand cultures with various fertiliser treatments. Weighed amounts of the crops were used to fertilise sugar beet, also grown in sand. The yield of sugar beet was closely related to the amount of nitrogen added in the green manure. Neither species nor fertiliser treatment to the green manure affected the growth of the sugar beet except when they influenced the nitrogen supplied to the sugar beet.

The value of green manuring in farming practice

The experimental evidence for the value of green manures for sugar beet indicates that they usually give quite large increases in yield. However, too few experiments have been made testing green manure in the presence of an adequate dressing of the major nutrients, particularly nitrogen. If this were done it seems likely that many of the reported increases from green manure would be very much smaller and often negligible. This was certainly the case at Broom's Barn, although at Woburn green manures appeared to have a beneficial effect on sugar beet which could not be explained in terms of the nutrients they supplied.

In deciding whether to use a green manure, the cost of seed and establishing the crop must be balanced against a saving in fertiliser for the sugar beet. Dyke[112] showed the practice could be economic at Woburn. In dry climates it must be remembered that the green crop uses soil reserves of moisture which might outweigh the theoretical benefits in yield of the sugar beet. Another consideration is the effect of cruciferous crops and relatives of sugar beet such as buckwheat (often grown as green manures) on sugar beet cyst eelworm (*Heterodera schactii*). These are hosts to the parasite and in England the sugar beet contract requires that they are not grown before sugar beet.

Chapter 9

Crop Rotations and Residual Value of Fertiliser

Sugar beet in the crop rotation

In most countries where sugar beet is cultivated successfully, it is grown in rotation with other crops. Where monoculture has been attempted, yields have usually declined rapidly, largely due to the multiplication of pests and diseases—particularly sugar beet cyst eelworm (*Heterodera schactii*). Weeds also become a problem where rotational cropping is not practised. Mainly for these reasons it seems likely that sugar beet will continue to be grown in some form of arable or ley/arable rotation. In farming practice it is therefore important, when considering the amount of fertiliser needed by sugar beet, to take into account the needs of the other crops in the rotation.

It is not possible to study any crop in isolation from the rest because residues of plants and unused fertiliser affect the amount of fertiliser needed by the following crop. Thus fertilisers applied to the crops grown before sugar beet affect the amount required by the sugar beet. The previous cropping also affects the amount of fertiliser required, for all crops take different quantities of nutrients from the soil and leave different nutrient residues. Sugar beet takes large quantities of some nutrients from the soil, which may affect the following crop. Sugar-beet tops are particularly rich in some elements and it is important to take this into account when the tops are removed from the field. Many experiments have been made to investigate how these factors affect the nutrient requirement of the sugar beet and how the nutrition of the sugar beet affects the following crop.

147

Residual effects of fertilisers and crop residues

FROM THE PREVIOUS CROP ON SUGAR BEET
Different farming systems affect the nitrogen fertiliser requirement
of sugar beet. Adams,[8] reporting experiments testing nitrogen on
commercial farms, noticed that where many cereal crops were grown,
the requirement increased (Table 71). On 22 of 27 fields where sugar

TABLE 71

RESPONSE TO NITROGEN FERTILISER IN RELATION TO
THE NUMBER OF PRECEDING CEREAL CROPS
(after Adams[8])

Number of preceding cereal crops	Number of experiments in which optimum nitrogen dressing for sugar beet was:	
	0·6 cwt/acre (75 kg/ha) N	more than 0·6 cwt/acre (75 kg/ha) N
0	6	0
1	16	5
2	7	10
>2	4	4

beet followed no or one cereal crop, 0·6 cwt/acre N was sufficient
on most of them, but where the sugar beet followed two or more
cereal crops more than 0·6 cwt/acre was often needed. Tinker[329] in
similar experiments also found that sugar beet after cereals needed
more nitrogen fertiliser than after other crops (Table 72), but there

TABLE 72

EFFECT OF PRECEDING CEREAL CROPPING ON THE
OPTIMUM NITROGEN DRESSING FOR SUGAR BEET
(after Tinker[329])

Number of preceding cereal crops	No. of fields	Average optimum N dressing (cwt/acre)	(kg/ha)
0	5	0·77	97
1	15	1·01	127
2	14	0·95	120
3	6	1·09	137

was no evidence that there was a greater need for nitrogen after two
or more cereal crops than after one. Boyd et al.[38] in 170 experiments
with nitrogen fertiliser found that the average responses to each
amount of nitrogen tested were somewhat less after potatoes than
the general average (Table 73).

TABLE 73
EFFECT OF PREVIOUS CROPPING WITH POTATOES ON SUGAR
YIELD RESPONSE TO NITROGEN
(after Boyd et al.[38])

	Sugar yield			
	Mean yield without N	Response to N fertiliser cwt/acre N		
		0·6–0·0	1·2–0·6	1·8–1·2
After potatoes	49·3	7·0	−1·6	−1·9 ⎫
All fields	45·2	8·6	1·2	−0·8 ⎭ cwt/acre
	Mean yield without N	Response to N fertiliser kg/ha N		
		75–0	150–75	225–150
After potatoes	6·19	0·88	−0·20	−0·24 ⎫
All fields	5·67	1·08	0·15	−0·10 ⎭ t/ha

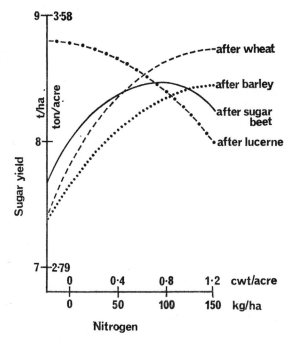

FIG. 24. Effect of previous cropping on response to nitrogen at Broom's Barn.[189]

Hull and Webb[189] grew sugar beet with four amounts of nitrogen fertiliser after sugar beet, wheat, barley and lucerne. Figure 24 shows that 0·4 cwt/acre N gave the greatest yield after lucerne, 0·8 cwt/acre after sugar beet and 1·2 cwt/acre after wheat and barley. Maximum yields after barley and sugar beet were less than those after wheat and lucerne. Draycott and Last[88] described experiments at Silsoe (Bedfordshire) and at Broom's Barn designed to investigate the effects of previous cropping and manuring on the nitrogen fertiliser needed by sugar beet. The Silsoe experiments tested barley (given 0·5 cwt/acre N) and potatoes (given 0·5 or 1·5 cwt/acre N) in the first year. In the second year the plots were sown to sugar beet and given 0, 0·6 or 1·2 cwt/acre N. Table 74 shows the yields of sugar

TABLE 74

EFFECTS OF PREVIOUS CROPPING AND MANURING ON SUGAR YIELD OF BEET GROWN WITH THREE AMOUNTS OF NITROGEN AT SILSOE

(after Draycott and Last[88])

	Sugar yield		
	N for sugar beet (cwt/acre)		
	0	0·6	1·2
Barley	59·6	68·9	66·9 ⎫
Potatoes, 0·5 cwt/acre N	64·4	67·7	68·6 ⎬ cwt/acre
Potatoes, 1·5 cwt/acre N	64·5	67·3	65·3 ⎭
	N for sugar beet (kg/ha)		
	0	75	150
Barley	7·48	8·65	8·40 ⎫
Potatoes, 63 kg/ha N	8·08	8·50	8·61 ⎬ t/ha
Potatoes, 189 kg/ha N	8·10	8·45	8·20 ⎭

as averages of three experiments. Sugar beet grown without nitrogen fertiliser after barley gave the smallest yield, but when given 0·6 cwt/acre N all the sugar-beet yields were similar and previous cropping and nitrogen manuring had very little effect.

At Broom's Barn, barley, winter wheat, barley under-sown with trefoil, potatoes (given two amounts of nitrogen) and a ley were grown in the first year. Table 75 shows the amounts of nitrogen given, the uptake and the residues of the first-year crops. In most cases the crops used slightly more nitrogen than the amount applied. There

TABLE 75

NITROGEN APPLIED, QUANTITY IN THE CROP AND NET GAIN OR
LOSS BY THE SOIL AT BROOM'S BARN
(after Draycott and Last [88])

	N applied	N uptake	N applied − N uptake
		(cwt/acre)	
Barley	0·5	0·64	−0·14
Winter wheat	0·8	0·78	+0·02
Barley, under-sown	0·5	0·68	−0·18 [+0·32]
Potatoes	0·5	0·68	−0·18
Potatoes	1·5	0·94	+0·56
Grass	0·5	0·61	−0·11
		(kg/ha)	
Barley	63	80	−18
Winter wheat	100	98	+3
Barley, under-sown	63	85	−23 [+40]
Potatoes	63	85	−23
Potatoes	188	118	+70
Grass	63	77	−14

[ ]—Including ploughed-down trefoil.

was a large residue of nitrogen after potatoes given 1·50 cwt/acre, and
the trefoil provided 0·51 cwt/acre N.

Table 76 shows the average yields of sugar beet grown in the
following year with four amounts of nitrogen. Yields were similar
after wheat and barley—about 1·00 cwt/acre N was needed for

TABLE 76

EFFECT OF PREVIOUS CROPPING AND MANURING ON YIELD OF
SUGAR BEET GROWN WITH FOUR AMOUNTS OF NITROGEN AT
BROOM'S BARN
(after Draycott and Last [88])

	N for sugar beet							
	cwt/acre				kg/ha			
	0	0·5	1·0	1·5	0	63	126	189
	Sugar yield							
	(cwt/acre)				(t/ha)			
Barley	55·9	62·3	65·2	63·8	7·02	7·82	8·18	8·01
Winter wheat	57·7	66·0	66·0	62·5	7·24	8·28	8·28	7·84
Barley, under-sown	60·0	62·9	62·8	61·0	7·53	7·89	7·88	7·66
Potatoes, 0·5 cwt/acre N	56·8	61·2	64·7	61·5	7·13	7·68	8·12	7·72
Potatoes, 1·5 cwt/acre N	56·6	65·6	61·7	61·3	7·10	8·23	7·74	7·69
Ley	47·7	62·0	64·8	60·4	5·99	7·78	8·13	7·58

maximum yield. On the undersown plots (where trefoil had been ploughed in) only 0·5 cwt/acre N of fresh fertiliser was needed for maximum yield. As at Silsoe, potatoes given much nitrogen decreased the nitrogen requirement of the sugar beet by 0·50 cwt/acre.

The three winters during the Broom's Barn experiments were relatively dry and there was a linear relationship ($r = -0.86$) between the nitrogen residue from the first-year crops and the amounts of fresh fertiliser nitrogen needed for maximum sugar yield. There was some indication that fresh nitrogen was slightly less effective than residual nitrogen.

RESIDUAL VALUE OF POTASSIUM

Warren and Johnston[357] reported experiments at Rothamsted on the Exhaustion Land where some plots had received no potassium manuring since 1856. (The Exhaustion Land site is a strip of $2\frac{1}{2}$ acres of arable land at the north end of Hoosfield, and derived its name from the unmanured cereal cropping which began in 1902 to measure the residual effects of manures that had been applied in previous experiments.)

The residues of potassium were worth 0·60 cwt/acre K_2O of new potassium fertiliser and in a somewhat similar experiment at Woburn the potassium was worth about the same on one site but only 0·17 cwt/acre K_2O on another. Both at Woburn and on the Exhaustion Land field the response curves to fresh potassium did not show a maximum with any amount of potassium tested, and yield on the enriched soil always exceeded that on the depleted soil.

In a later report, Johnston et al.[195] found that potassium residues increased sugar yield by 10 cwt/acre on the Exhaustion Land and at Woburn by between 5 and 10 cwt/acre, but the yield of sugar was the same on depleted and enriched soils provided a large dressing of new potassium was given. The residues on the Exhaustion Land could not be valued in terms of a new dressing of potassium but at Woburn they were worth between 0·67 and 0·76 cwt/acre K_2O as fresh potassium fertiliser.

The manurial value of sugar-beet tops

Sugar-beet tops are particularly rich in nitrogen and cations but surprisingly few field experiments have been made to measure their manurial value to following crops. Pizer[277] drew attention to the organic matter they provide, particularly on poor, sandy soils. In many stockless rotations without leys and where cereal straw is burnt, sugar-beet tops and cereal roots are the only crops which return any quantity of organic matter. He estimated that sugar beet leaves little organic matter residue in the soil as roots but $1\frac{1}{2}$–3 tons/acre dry matter as leaves and crowns.

In France, Crohain and Rixhon[70] made a thorough examination of the value of sugar-beet tops in terms of nitrogen fertiliser on three succeeding cereal crops. In the first year after ploughing down, the tops were equivalent to 0·28–0·36 cwt/acre N. In the second cereal crop the tops increased yield equivalent to 0·08–0·10 cwt/acre N, but in the third crop it was difficult to find any residual effect. It is interesting to compare these values with the amount of nitrogen in the tops as determined by analysis before ploughing-down; the total nitrogen value in this case was 0·91 cwt/acre.

TABLE 77

YIELD OF BARLEY GRAIN AFTER SUGAR BEET: MEANS OF 3 EXPERIMENTS
(after Widdowson[372])

Sugar-beet tops	Barley grain			
	N dressing for barley (cwt/acre)			
	0	0·33	0·66	1·00
Removed	26·4	35·1	38·2	37·6 ⎫
Ploughed-in	29·9	36·0	38·8	37·3 ⎬ cwt/acre
	+3·5	+0·9	+0·6	−0·3 ⎭
	N dressing for barley (kg/ha)			
	0	41	83	126
Removed	3·31	4·41	4·79	4·72 ⎫
Ploughed-in	3·75	4·52	4·87	4·68 ⎬ t/ha
	+0·44	+0·11	+0·08	−0·04 ⎭

Widdowson[372] made similar experiments at Broom's Barn. Sugar-beet tops were either ploughed in or removed and barley grown in the following year with four amounts of nitrogen. Table 77 summarises the results. Where no nitrogen was given to the barley, the tops increased yield by 3·5 cwt/acre grain. However, only a small dressing of nitrogen fertiliser was needed to give the same effect. The ploughed-in tops were 'worth' about 0·15 cwt/acre N fertiliser given to the barley, and this value was remarkably similar in all three experiments.

Response to fertilisers by sugar beet in long-term experiments

WOBURN LEY–ARABLE 1956–67

As the name suggests, the main aim of this experiment was to assess the value of leys in an arable farming system but some of the fertiliser tests included have helped to elucidate some important aspects of the long-term effects of fertilisers on sugar-beet yields. The experiment was begun in 1937 and over a five-year cropping cycle compared the effect on the yield of two arable test crops of three years of ley or three years of arable cropping.

TABLE 78

MEAN YIELDS AND RESPONSES TO FARMYARD MANURE AND TO NITROGEN AND POTASSIUM FERTILISER BY SUGAR BEET IN THE WOBURN LEY–ARABLE EXPERIMENT 1956–61
(after Boyd[36])

	Three year grazed ley	Three year lucerne for hay	Arable (including seeds hay)	Arable (roots)
			Sugar yield	
		(cwt/acre)		
Mean yield	56·6	52·3	48·6	52·5
Response to:				
Farmyard manure (15 ton/acre)	6·8	9·4	12·5	14·6
Extra N*	−2·8	−3·0	−2·4	0·1
Extra K*	2·7	2·3	2·2	−0·4
		(t/ha)		
Mean yield	7·10	6·56	6·10	6·59
Response to:				
Farmyard manure (38 t/ha)	0·85	1·18	1·57	1·83
Extra N*	−0·35	−0·38	−0·30	0·01
Extra K*	0·34	0·29	0·28	−0·05

N*: Basal dressing 0·72 cwt/acre (90 kg/ha) plus a further 0·72 cwt/acre (90 kg/ha).
K*: Basal dressing 0·90 cwt/acre K_2O (94 kg/ha K) plus a further 0·90 cwt/acre K_2O (94 kg/ha K).

Until 1956, potatoes were the first test crop and barley the second.[239] The cropping scheme was changed for the period 1956–67 and sugar beet was taken in place of potatoes as the first test crop.[36] Early results indicated that potassium, nitrogen and possibly other elements such as magnesium were being depleted by the arable rotation. When the scheme was changed to include sugar beet, plots were split to test more nitrogen and potassium.

Table 78 shows that there were large effects of rotation and large responses to farmyard manure. On the continuous arable sequence, farmyard manure increased yields of sugar by 18 cwt/acre (more than

40%), whereas on the ley sequence the increase was only 6 cwt/acre or 10%. There was little increase in yield from the extra nitrogen and potassium applied, an unexpected result. Soil analyses showed that the crop *should* have responded to potassium. Warren and Johnston[358] on a nearby experiment showed that the lack of response was due to the distribution of the potassium in the soil. Where it was dug in, the crop responded; where it was broadcast on the surface, there was little response. Thus the yields of sugar were larger with farmyard manure than with fertilisers because the farmyard manure contained large amounts of potassium that had been ploughed-in.

Table 79 shows the mean sugar yields for 1962–64, during which years the potassium status of plots in each rotation and of plots

TABLE 79

MEAN YIELDS AND RESPONSES TO FARMYARD MANURE AND TO NITROGEN AND POTASSIUM FERTILISER BY SUGAR BEET IN THE WOBURN LEY-ARABLE EXPERIMENT 1962–64
(after Boyd[36])

	Three year grazed ley	Three year lucerne for hay	Arable (including seeds hay)	Arable (roots)
			Sugar yield	
		(cwt/acre)		
Mean yield	60·6	58·8	54·2	61·0
Response to:				
Farmyard manure	0·9	3·7	4·3	5·7
Extra N	−2·4	−1·8	1·8	0·3
Extra K	0·9	2·3	0·7	−1·1
		(t/ha)		
Mean yield	7·61	7·38	6·80	7·66
Response to:				
Farmyard manure	0·11	0·46	0·54	0·72
Extra N	−0·30	−0·23	0·23	0·04
Extra K	0·11	0·29	0·09	−0·14

with and without farmyard manure were equalised by giving large corrective dressings of fertiliser. The effect was to decrease response to farmyard manure from more than 10 to less than 4 cwt/acre sugar. For all rotations except lucerne, responses to potassium were less than in previous years. Responses to N differed between rotations but it was not possible to indicate precisely the optimal nitrogen dressing after each rotation. From 1965 to 1967, four amounts of nitrogen were tested and sugar beet in the arable rotations needed most, the optima depending on whether or not farmyard manure was used.

Thus at Woburn these 12 years of results with sugar beet indicate that the early large effects of farmyard manure can be explained mainly in terms of response to nutrients, but there was still a small effect of farmyard manure not obtainable by fertilisers alone. Another experiment at Woburn also emphasised the importance of thorough mixing of the fertiliser dressing with the soil for maximum response, so that it is not concentrated in the surface soil where it may be largely unavailable in dry periods.

SWEDISH LEY–ARABLE EXPERIMENT

Agerberg[9] has reported on a similar ley–arable experiment in Sweden. Three four-course rotation systems were tested: A—sugar beet (given 8 ton/acre farmyard manure), barley, ley and winter wheat; B and C—sugar beet, barley, winter rape and winter wheat. In B, the straw and sugar-beet tops were ploughed-in, and in C the straw was burned and the tops carted off. Four rates of compound fertiliser were given during the first rotation and the plots were split for nitrogen dressings during the second rotation.

Yields during the first eight years showed no marked advantage from the ley system although analysis indicated that the three systems

TABLE 80

EFFECT OF THREE ROTATION SYSTEMS AND OF NITROGEN
FERTILISER ON YIELDS OF SUGAR BEET IN SWEDEN
(after Agerberg[9])

| | | | Rotation | |
| | | A | B | C |
			Yield of roots	
			(ton/acre)	
1st Crop		22·5	23·0	23·6
2nd Crop				
N dressing: (cwt/acre)	0	16·4	14·3	14·5
	0·32	16·9	16·2	17·5
	0·64	18·4	18·2	17·3
	0·96	18·0	17·5	18·5
			(t/ha)	
1st Crop		56·4	57·9	59·4
2nd Crop				
N dressing: (kg/ha)	0	41·2	35·9	36·3
	40	42·5	40·6	43·9
	80	46·1	45·6	43·4
	120	45·2	43·9	46·4

changed the amount of organic matter in the soil. In system A it increased but in B, and more especially in C, the organic matter decreased. There was a tendency for greatest losses where least fertiliser was given. Table 80 shows the yields of roots for the two sugar-beet crops. The rotation systems affected the amount of nitrogen fertiliser required for maximum yield and the magnitude of the response to nitrogen, but given the correct nitrogen dressing the yields were similar from all three systems.

BARNFIELD—CONTINUOUS SUGAR BEET
Few field experiments in England or indeed anywhere in the world have tested continuous sugar-beet cultivation over a long period. Barnfield experiment at Rothamsted probably comes nearest to a sugar-beet equivalent of the Broadbalk wheat experiment. Lawes and Gilbert started a manurial experiment on mangolds on Barnfield in 1876. Mangolds were grown continuously until 1959 except for a few years when the crop failed. From 1946 to 1959 a part of each plot (the same part each year) was cropped with sugar beet. The results up to 1894 were summarised by Lawes and Gilbert in 1895 and reported in detail by Hall.[152] Watson and Russell[362] examined the results up to 1940, and Warren and Johnston[360] up to 1959.

The experiment tested various nitrogen fertilisers, phosphorus, potassium, sodium and magnesium fertilisers, rape cake and farmyard manure. Yields of mangolds up to 1945 gave much interesting information, recently summarised by Cooke,[66] but these results will not be dealt with here. Table 81 shows the average yields of sugar beet in 1946–59 (except for two years when the crop failed). Ammonium and nitrate nitrogen gave the same increase in sugar-beet root yield (5 tons/acre) in the presence of phosphorus and potassium or phosphorus and sodium. Potassium and sodium each increased yield by 2 tons/acre where the nitrogen was given as ammonium sulphate, but neither had any effect where sodium nitrate was used. Fertilisers out-yielded farmyard manure by 1–2 tons/acre.

Barnfield also provides unique data relating to nutrient uptake by mangolds and sugar beet; also corresponding changes in the quantities of nutrients in soil brought about by long continued cropping, although detailed chemical analyses of crops and soils were only made in the later years of the experiment.[360] Uptake of nutrients by sugar beet on Barnfield are given on page 39 and changes in soil nutrient status on page 69.

SAXMUNDHAM—NORFOLK FOUR-COURSE
A long-term rotation experiment was begun at Saxmundham in 1899 with the Norfolk four-course system of wheat, barley, mangolds and

TABLE 81

YIELDS OF SUGAR-BEET ROOTS FROM BARNFIELD: MEANS 1946–59
(after Warren and Johnston[360])

	No nitrogen	Ammonium sulphate	Sodium nitrate	Rape cake	Rape cake + ammonium sulphate
			(ton/acre)		
No P or K	1·5	4·2	5·0	5·6	6·4
P	1·9	5·0	6·7	6·9	7·2
P K	1·6	6·6	6·2	8·2	9·5
P Na Mg	1·8	7·2	7·2	7·7	9·0
P K Na Mg	1·8	7·2	8·0	9·1	10·3
FYM	6·2	11·5	11·1	11·4	11·4
FYM + P K	5·9	8·6	9·9	9·8	10·3
			(t/ha)		
No P or K	3·8	10·6	12·6	14·1	16·1
P	4·8	12·6	16·8	17·3	18·1
P K	4·0	16·6	15·6	20·6	23·9
P Na Mg	4·5	18·1	18·1	19·3	22·6
P K Na Mg	4·5	18·1	20·1	22·9	25·9
FYM	15·6	28·9	27·9	28·6	28·6
FYM + P K	14·8	21·6	24·9	24·6	25·9

beans and peas. In 1965 sugar beet was grown in place of mangolds and the four-course system continued until 1969. From 1956 to 1964 sugar beet and mangolds were grown side by side on half plots. From 1899–1961 eight fertiliser treatments were applied for each crop: none and 0·30 cwt/acre N, none and 0·30 cwt/acre P_2O_5 and none and 0·50 cwt/acre K_2O, in factorial combination. Farmyard manure (6 tons/acre) and bonemeal (4 cwt/acre) were also tested. In 1962 the fertiliser treatments were changed to test dressings near to current commercial practice. The experiment was described by Trist and Boyd[338] for the period 1899–1961 and by Williams and Cooke[375] for the period 1964–69.

Table 82 shows the average yields of mangolds (1906–61) and sugar beet (1964–69). Response to phosphate was outstanding, for yields were doubled. Response to nitrogen was less than to phosphorus and very small unless phosphorus was also given. Yields of mangolds were less with farmyard manure than with a full dressing of fertilisers, but yields of sugar beet were greater with a full dressing of fertiliser than with farmyard manure.

WOBURN REFERENCE PLOT EXPERIMENT

An experiment at Woburn was begun in 1960 on a field where arable crops had been grown for many years. The first five years of cropping was described by Widdowson and Penny.[370] Nitrogen, phosphorus and potassium fertilisers and farmyard manure were tested during a five-course rotation of barley, ley, potatoes, oats and sugar beet.

Sugar beet yielded the most and barley the least dry matter, both with and without fertiliser and manure. All crops responded greatly to nitrogen fertiliser but only slightly to phosphorus; response to potassium increased each year. The amount of nitrogen, phosphorus and potassium in the crops was measured, and sugar beet contained most nitrogen (0·7 cwt/acre) and oats and barley least (0·2 cwt/acre). The ley and sugar beet contained most phosphorus (0·4 cwt/acre P_2O_5) and barley least (0·1 cwt/acre). The ley and sugar beet also removed most potassium (1·5 cwt/acre K_2O) and barley least (0·2 cwt/acre).

TABLE 82

AVERAGE YIELDS OF MANGOLD (1906–61) AND SUGAR-BEET (1964–69) ROOTS IN SAXMUNDHAM ROTATION I EXPERIMENT
(after Trist and Boyd[338] and Williams and Cooke[375])

| Treatment | Root yield | | | |
| | (ton/acre) | | (t/ha) | |
	Mangolds	Sugar beet	Mangolds	Sugar beet
None	4·2	2·7	10·6	6·8
K	4·2	2·7	10·6	6·8
N	4·9	4·0	12·3	10·0
N K	5·6	3·4	14·1	8·5
P	11·3	5·7	28·4	14·3
P K	12·1	6·0	30·4	15·1
N P	17·0	9·3	42·7	23·4
N P K	18·0	9·2	45·2	23·1
FYM	17·0	12·9	42·7	32·4
Bonemeal	11·6	6·4	29·1	16·1

The amount of each element apparently recovered from fertiliser was also measured.[371] Sugar beet recovered most nitrogen (67%) and potatoes least (35%). The crops recovered only a little phosphorus (2–12%) but much potassium—the ley most (57%) and barley least (9%). Over the five-year period the largest losses from the soil were 1·8 cwt/acre N, 1·3 cwt/acre P_2O_5 and 4·3 cwt/acre K_2O, and the largest gains to the soil (from manuring) 4·5 cwt/acre N, 5·5 cwt/acre P_2O_5 and 5·4 cwt/acre K_2O.

SIX-COURSE ROTATION EXPERIMENTS, WOBURN AND ROTHAMSTED

Nitrogen, phosphorus and potassium fertilisers were tested on a six-course rotation of crops at Woburn and Rothamsted. Results for the period 1931–55 were reported by Yates and Patterson.[383] The 15 fertiliser treatments were applied so that each plot received every treatment once in 15 years. The six crops grown were wheat, barley, rye, potatoes, sugar beet and clover. Responses were estimated (by the method of Crowther and Yates)[71] to standard dressings of 0·25 cwt/acre N, 0·5 cwt/acre P_2O_5 and 0·5 cwt/acre K_2O. Table 83

TABLE 83

ESTIMATES OF THE RESPONSES BY SUGAR BEET TO STANDARD DRESSINGS OF FERTILISERS IN THE SIX-COURSE ROTATION EXPERIMENTS AT ROTHAMSTED AND WOBURN 1931–55
(after Yates and Patterson[383])

	N	Response to P Sugar yield	K
		(cwt/acre)	
Rothamsted	2·5	−0·8	0·5
Woburn	4·5	−0·4	2·3
		(t/ha)	
Rothamsted	0·31	−1·00	0·06
Woburn	0·57	−0·50	0·29

shows the responses for sugar beet. There was a large response to nitrogen at both stations, particularly at Woburn, but average response to phosphorus was negative at Rothamsted and Woburn. Potassium increased yield considerably at Woburn but gave little increase at Rothamsted.

One of the main aims of these experiments was to relate the response to fertilisers and weather. This was not very successful, due partly to the small size of the response and partly to faults in the design of the experiments. There was some evidence, however, that the sugar beet responded best to nitrogen in wet summers, both at Rothamsted and Woburn.

Conclusions

SHORT-TERM RESIDUES

The amount of fertiliser needed by sugar beet (or any crop) is influenced by the residues from the previous crop. Both unused fertiliser and crop residues provide part of the nutrients required and it is worthwhile deducting this amount from the dressing which would otherwise be used. Table 84 shows the residues of phosphorus and

TABLE 84

THE VALUE OF FERTILISER AND CROP RESIDUES

	Sugar beet: tops removed	Sugar beet: ploughed-in	Barley: grain plus straw	Wheat: grain plus straw	Potatoes: tubers	Ley: cut and removed	Beans: grain plus straw
	(cwt/acre)						
Fertiliser dressing:[a]							
Phosphorus (P_2O_5)	0·5	0·5	0·4	0·4	1·3	0·5	0
Potassium (K_2O)	1·0	1·0	0·3	0·3	1·0	1·0	0·3
Offtake:[b]							
Phosphorus (P_2O_5)	0·5	0·3	0·2	0·3	0·3	0·4	0·3
Potassium (K_2O)	1·8	0·7	0·3	0·5	1·0	1·5	0·4
Residue:							
Phosphorus (P_2O_5)	0	+0·2	+0·2	+0·1	+1·0	+0·1	−0·3
Potassium (K_2O)	−0·8	+0·3	0	−0·2	0	−0·5	−0·1
Nitrogen (N)	0	+0·15	0	0	+0·6	+0·2	+0·4
	(kg/ha)						
Fertiliser dressing:[a]							
Phosphorus (P)	27	27	22	22	71	27	0
Potassium (K)	104	104	31	31	104	104	31
Offtake:[b]							
Phosphorus (P)	27	16	11	16	16	22	16
Potassium (K)	188	73	31	52	104	156	42
Residue:							
Phosphorus (P)	0	+11	+11	+6	+60	+6	−16
Potassium (K)	−83	+31	0	−21	0	−52	−10
Nitrogen (N)	0	+19	0	0	+75	+25	+50

[a] Recommended dressing for soils with 16–25 ppm P and 100–200 ppm K (ADAS index 2).
[b] Means of six years of crops at Broom's Barn given a commercial dressing of fertilisers.

potassium fertiliser as the difference between fertiliser dressing and offtake of six common crops. As there are other losses in addition to that removed in crops, the small residues shown for most crops may be ignored for one crop but may be considerable over a rotation. Potatoes leave a large residue, and sugar beet (where tops are removed) and grass deplete soil potassium reserves.

Residual values for nitrogen cannot be calculated reliably by this 'nutrient balance sheet' method used for phosphorus and potassium. The values for nitrogen given in Table 84 were obtained at Broom's Barn; for sugar beet the residues were measured in the following barley and for the other crops, in the following sugar beet. Although sugar-beet tops contained 1·0 cwt/acre N they were only 'worth' 0·15 cwt/acre of fresh fertiliser N to the barley. Barley and wheat left little residual nitrogen for sugar beet but potatoes were 'worth' 0·6 cwt/acre, a cut ley 0·2 cwt/acre and field beans 0·4 cwt/acre of fresh fertiliser N. Thus when sugar beet follows these crops the fertiliser dressing can safely be decreased by these amounts of nitrogen fertiliser.

LONG-TERM RESIDUES

Sugar-beet growers in many countries now give more fertiliser than experiments show is sufficient for maximum yield, in the belief that unused fertiliser increases soil fertility, with eventual increased yields of sugar beet and other crops. Others include crops in the rotation to maintain or increase soil organic matter, believing that this, too, leads to increased yields. The value of these 'long-term' residues can only be measured in expensive, detailed experiments over many years, few of which have been made.

The ley–arable experiments described in previous pages give little support to the suggestion that improved yields result from increased soil organic matter from leys or farmyard manure. Where sufficient fertiliser was given in wholly arable rotations, yields were equal to those after leys or those grown with farmyard manure (e.g. Table 78). In a long-term experiment at Broom's Barn, yields of sugar beet are being compared when grown in five contrasting rotations: sugar beet every year; once in three years with two barley crops; once in six years with five barley crops; once in three years with a two-year ley, and once in three years with beans and potatoes. During the first six years of this experiment the yields of sugar beet in all rotations did not differ significantly.[105]

Another experiment explores the value of doubling the amounts of fertiliser nitrogen, phosphorus and potassium recommended for Broom's Barn soil for each crop in a three-course rotation of sugar beet, winter wheat and barley. Results of the first six years show that, whereas the recommended dressings of nitrogen, phosphorus and potassium balance the amounts removed in the crops, large residues accumulated with the double dressing. Soil analysis confirms substantial increases in available phosphorus and potassium. The experiment so far provides little evidence that the recommended dressings need increasing when the yields during six years of cropping are considered.[104]

Chapter 10

Soil Physical Conditions

In addition to being present in sufficient quantity, nutrients and water must be accessible to plant roots and the soil must provide a suitable environment for the roots to grow. The sugar-beet crop, particularly just after sowing, is very sensitive to soil physical conditions and to establish and yield well, sugar-beet roots must be able to extend rapidly. Soil *texture* is defined by the relative amounts of clay, silt and sand particles, and greatly influences soil physical conditions but there is no practical way of altering soil texture. However, soil *structure*, or the aggregation of particles into larger 'crumbs', *is* influenced by weather and cultural treatments but it is difficult to define and measure. The growth of sugar beet in relation to practices which alter soil physical properties has been investigated in the experiments below.

Previous cropping

Robertson[288] in Michigan, investigated the effect of several systems of crop rotation on soil structure in relation to production of sugar beet. In addition to measuring yields, some physical characteristics of the soil were studied. Grass, legumes and leys increased the volume of soil pores and the yield of sugar beet. At the time of planting, the soil on the plots where a legume was grown in the previous year was in a more stable state of aggregation than soil from other plots. Percentage pore space, both total and non-capillary, was increased by growing legumes before the sugar beet. Differences in pore space were greater during the latter part of the growing season than at the beginning of the season. Penetrometer studies showed that the hard crust formed on the surface of the soil of control plots did not form to any appreciable extent where legumes had been grown. Since greater yields and improved soil structure resulted from different cropping systems, the author concluded that some of the increase in yield was simply due to improved soil structure.

Sodium

Many sugar-beet growers avoid using sodium fertiliser for the crop in the belief that a dressing of the element damages soil structure but there is no experimental evidence that this is so. Giving 3 to 4 cwt/acre NaCl in several hundred annual experiments throughout Britain consistently increased sugar-beet yield without visible damage to soil structure. Similarly, large applications in the 2-Course Experiment at Rothamsted had no effect on soil structure. On silt and clay soils, the increase in yield is usually less than on sand and loam soils because silts and clays generally supply large amounts of potassium, replacing some of the need for sodium (*see* page 78).

In naturally occurring saline soils in arid climates and in soils in Great Britain which have been flooded by sea water, the sodium disperses soil clays, which damages soil structure and increases bulk density. When dry, sodium-dispersed soils are consolidated and difficult to distinguish from ones which have been compacted mechanically. Sodium-dispersed clay soil also has a tendency to crack as it dries, but compacted soil shrinks little.

Compaction

Soil compaction caused by heavy machinery, by ill-timed cultivations or by natural soil processes results in increased soil density. This decreases the volume of large pores and increases the proportion of fine pores. Severe compaction causes mechanical impedance to root growth and even slight compaction may adversely affect plant growth by causing anaerobic conditions to develop. The availability of nutrients may also be diminished by restricted root development and by anaerobic soil conditions.

Root Crop

Experiments at Broom's Barn and Saxmundham have investigated the effect of soil compaction on yield, seedling emergence and fertiliser requirement of sugar beet.[91] Some seedbeds were compacted in winter, others in spring and others prepared with a minimum of compaction. Each was tested with 0·6, 1·2 and 1·8 cwt/acre N. Compaction decreased seedling populations in four experiments but increased it in one when the weather was dry while the seeds were germinating. However, in every experiment compaction greatly decreased yield of roots and sugar. It also interacted with the fertiliser treatment, increasing the amount of nitrogen required for maximum yield. On average, 0·6 cwt/acre N gave the greatest yield without

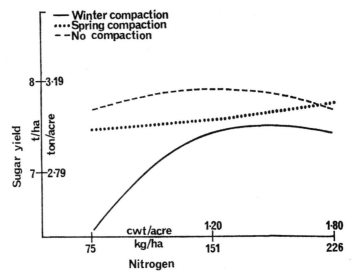

FIG. 25.   Effect of soil compaction on response to fertiliser.[91]

compaction and 1·2 cwt/acre N with compaction (Fig. 25). The decreased requirement of nitrogen can be explained in terms of restricted root development and decreased available soil nitrogen, resulting from anaerobic conditions which developed following compaction.

SEED CROP
Pendleton[268] investigated the effect of compacting plots of sandy loam soil where sugar-beet seed crop was to be grown. Deep tillage and heavy applications of farmyard manure decreased soil compaction. This resulted in roots of improved shape and more rapid growth of the seed crop in the autumn, but seed yields were improved little. All forms of cultivation and applications which decreased compaction improved root shape. Compacting the soil in the greenhouse to a porosity of 3·5% (i.e. 3·5% of the soil volume taken up by air) restricted all development of sugar-beet roots. With a porosity of 14% the root distribution and development was much improved and better than soil with a porosity of 6·5%. In a silt loam soil, sugar-beet root development was restricted when the porosity was 11·7% but was rapid at 18%.

RED BEET
Kubota and Williams[207] studied the effect of changes in soil density and porosity on germination, establishment and yield of red beet.

Two degrees of compaction, heavy—with a flat-tyred vehicle wheel, and light—with a ring-roller, were given to the seedbed after sowing red beet on three contrasting soils. Changes in pore space of the soils and the responses of the crop to the changes in physical properties were measured. The experiments were at Rothamsted, on a heavy clay loam which had long been under arable cultivation and contained little organic matter, on another field at Rothamsted after eight years of grass, and at Woburn on a light sandy loam long in arable cultivation and containing little organic matter.

The contributions of the large-sized pores to the total pore space in 1-in diameter clods was lessened by compaction, whereas the small-sized pores (less than 0·04 mm diameter) in the 1–2 mm aggregates were not affected. Effects of compaction generally became less with depth, and were negligible 8 to 10 in deep. The heavy soils compacted more than the light soil. The heavy soil where the ley had grown showed very little effect from compaction at 3 to 6-in depth. On the light soil, compaction increased the tendency to form a hard surface pan which severely restricted root penetration, but on the heavy soils surface cracking diminished the adverse conditions for crop growth caused by compaction. Light compression with a ring-roller did not depress yield of beet on the heavy soils but it did on the light soil. Heavy compaction with the vehicle wheel greatly reduced yield on the light soil, less on the arable soil and least on the heavy soil following a long ley.

The concentration of nitrogen, phosphorus, potassium and manganese in the crops at harvest were not consistently related to the compaction treatments. Compaction lessened the ratio of phosphorus-to-nitrogen uptake by the crops on the light soil but effects on crop uptake from the heavy soils were much smaller. The effects of soil compaction on the germination, growth and yield of the beet were much greater than on barley. The authors conclude that red beet is very sensitive to structural soil conditions whereas barley is relatively insensitive.

Deep ploughing and subsoiling

Russell[295] studied the effects of ploughing to more than 12 in deep and of subsoiling to a depth of about 18 in on sugar-beet yield in relation to response to phosphorus and potassium fertiliser. The sugar-beet yield was consistently increased by the deep tillage, and fertiliser phosphorus and potassium given before ploughing increased yield by $\frac{1}{2}$ to 1 ton/acre of roots compared with fertiliser applied on the seedbed, but depth of ploughing did not affect the response to fertiliser. Hull and Webb[189] investigated the effects of subsoiling with tines 4 ft apart and 20 in deep on response by sugar beet and other crops to fertilisers at Broom's Barn. Subsoiling increased root yield by about 0·6 ton/acre but did not affect the response to fertiliser nor the amount of fertiliser required for maximum yield.

Sugar-beet growing without ploughing

Hagan[148] discussed some of the advantages of growing sugar beet without ploughing on sandy soil subject to wind erosion and described some of the techniques employed. Establishing the crop in this way conserved seedbed moisture, prevented erosion, eased harvesting and resulted in moderate yields. However, no comparative yields were available for the crop on ploughed soil. In Holland, Kupers and Ellen[208] grew sugar beet on ploughed and unploughed plots and tested four amounts of nitrogen (Table 85). Ploughing

TABLE 85

RESPONSE TO NITROGEN BY SUGAR BEET ON PLOUGHED AND UNPLOUGHED SOIL
(after Kupers and Ellen[208])

		N dressing						
	(cwt/acre)				(kg/ha)		
	0	0·48	0·96	1·44	0	60	120	180
				Total dry matter yield				
	(ton/acre)				(t/ha)		
Unploughed	5·2	6·5	7·0	7·6	13·0	16·4	17·6	19·0
Ploughed	6·1	7·9	8·6	8·6	15·3	19·8	21·7	21·5

improved the yield and also decreased the amount of nitrogen needed for maximum yield. It also increased the proportion of dry matter partitioned to the roots, and during hot spells in the summer the crop on the unploughed plots often wilted, suggesting that ploughing had also improved the root system and the water uptake by the crop.

Sugar-beet growing on soils of unstable structure

Cooke and Williams[67] have discussed some of the problems of growing sugar beet on the difficult Saxmundham soil. There is only a small range in moisture content where this badly-drained sandy clay is not too wet to work, and not too dry for large aggregates to be broken by cultivating. Even with good under-drains, such soils do not weather to give good seedbeds, for fine crumbs overlie massive clods that remain from ploughing. This is a most important property of soil for successful arable cropping—clods formed by ploughing, cultivating or traffic must break up as they dry and, after rewetting, to form a tilth naturally. Longden[227] investigated the value of several soil conditioners on sugar-beet establishment and yield at Saxmundham. Farmyard manure, peat, 'Lytag', plastic wastes and 'Krilium' all had no effect on seedling emergence or their weight. Yields at harvest were not affected by conditioners though they made the roots more fangy.

Optimum soil moisture for emergence

Stout et al.[323] made a laboratory study to determine the optimum range of soil moisture and compaction for germination and emergence of sugar-beet seedlings. The results showed that soil moistures from 12 to 21% were satisfactory, whilst soil moistures from 16 to 22% were even better. In soil near the critically small moisture content, the addition of small quantities of water above the seeds increased the emergence considerably. The optimum compaction pressure was from 2 to 5 lb/in^2. Emergence was decreased when no pressure was applied and also with pressures about 5 lb/in^2, and the effect of compaction varied with the soil moisture content. In soil at 12 to 16% moisture, compaction pressure above 5 lb/in^2 reduced emergence, while in soil at 21% moisture, compaction pressures ranging from 2 to 30 lb/in^2 did not affect emergence.

Optimum porosity

Baver and Farnsworth[18] in Ohio, USA, investigated the relationship between the number of plants lost during the interval from singling time to harvest as influenced by the aeration of the soil. Losses of nearly 50% were observed where the non-capillary porosity was below 2% by volume (Fig. 26). They also found that the shape of the sugar-beet tap root was determined to a great extent by the porosity of the soil. Soils with poor aeration produced short stubby beet with many auxiliary roots whereas well-aerated soils produced long tapering roots. As a result, soils with non-capillary porosities exceeding 7–10% produced greater yields with larger sugar percentages. The maximum benefit from fertiliser applications were only obtained on soils where the structure allowed adequate aeration for the crop.

In Minnesota, USA, Blake *et al.*[23] compacted soil before growing sugar beet. The number of fangy roots was increased and the sugar

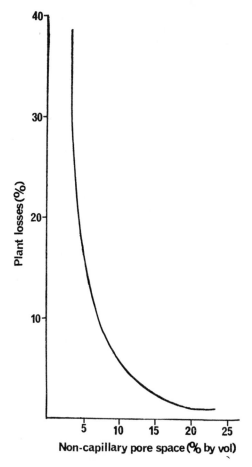

FIG. 26.   Non-capillary pore space in soil and the percentage of plants lost between singling and harvest.[18]

percentage decreased by compaction. They suggest a limiting non-capillary porosity for sugar beet of between 10 and 15%. Below 10%, gaseous diffusion approached zero and root damage often occurred even when the soil porosity dropped below this limiting value for only short periods.

Damage to soil structure caused by harvesting

Yields of cereals and other crops are frequently depressed by the damage caused by harvesting sugar beet. This is particularly noticeable when harvesting the crop on heavy soils in wet conditions. Cooke and Williams[67] drew attention to the poor yields of barley at Saxmundham in 1967 and 1969, partly due to harvesting in difficult conditions and partly to ploughing when the soil was wet. Experience at Broom's Barn is similar, for cereals do not yield well on the heavier soils if sugar-beet tops and wet, slaked soil are ploughed down. An ill-drained layer of structureless soil and decomposing tops becomes anaerobic during wet weather and cereal roots grow poorly.

Soil 'conditioners' and sugar-beet growth

Several soil conditioners have been tested for sugar beet grown on problem soils. Some conditioners such as 'Krilium' (a hydrolysed polyacryonitrile) and calcium sulphate act by changing the chemistry of soil, for example by flocculating clay particles. Others, such as peat and coconut fibre, alter the density and other related physical characteristics of soil. Large dressings of animal and plant residues like farmyard manure act chemically and physically in soil improvement (*see* Chapter 8).

IN BRITAIN
Hoyt[184] compared 'Krilium' (1 000 lb/acre), farmyard manure (62 ton/acre of dry matter), peat (18 ton/acre of dry matter) and coconut fibre (3 ton/acre dry matter) applied for sugar beet on a soil where roots of previous crops grew badly. 'Krilium' and farmyard manure increased yield of sugar-beet roots by 28 and 34 % respectively. Peat gave a smaller increase and coconut fibre none. The increases were most probably due to improved soil structure, for sufficient water and nutrients were given for maximum yield.

IN THE USA AND CANADA
Baird *et al.*[15] tested various chemical additives to clay soils of Dakota, USA, in an attempt to improve seedling emergence. Emergence was improved in two out of three experiments because the chemicals decreased surface capping. Also in USA, Smith[316] found that 2 000 lb/acre Krilium increased seedling growth by improving soil aeration. Bolton and Aylesworth[30] applied conditioner to 'problem' soils in Ontario, Canada. It improved the physical properties of the soil, which resulted in increased root and sugar yield. Williams *et al.*[377]

found that ploughing-in bulky organic residues before sugar beet in California increased the infiltration rate of irrigation water and increased yield. Compaction of the soil with tractors decreased yield.

Conclusions

Sugar beet is one of the most sensitive of common agricultural crops to soil physical conditions, particularly during germination and in the seedling stage. Good brairds depend on maximum germination, survival and vigorous seedling growth, all of which are influenced by soil physical conditions. Surprisingly few measurements have been made of the soil physical properties which govern whether a seedbed is good or bad, the main reason doubtless being the complexity of the problem. Soil texture, the nature of the clay fraction, organic matter, water content, porosity and other factors interact with each other and to a greater or lesser extent influence germination and growth of the plant.

Reports from the USA indicate that the water content of soil at germination should be between 16 and 22% (*i.e.* slightly less than field capacity) for maximum emergence, although germination can be satisfactory in soil with as little as 12% water. The percentage of the soil volume taken up by air (*i.e.* the porosity) is also important and American authors suggest that this should not be less than 12% for successful sugar-beet cultivation. This is particularly important when the seed is germinating and the plant is emerging. The bulk density of soil greatly affects the growth of roots and for sugar beet the limiting density is about 1·6 g/cm^3. Increasing bulk density by compaction causes physical impedance to root extension and decreases the volume explored by roots. There are several consequences of compaction, the main one being that the amount of fertiliser needed is increased (*e.g.* by about 0·60 cwt/acre N) for plants obtain less from residues. The crop on compacted soil is also more responsive to irrigation for less *soil* water is available to it as a result of the restricted root development. Recent experiments at Broom's Barn have shown that when sugar beet is grown only between tractor wheels (on beds or ridges) with the minimum of mechanical compaction, yields are increased compared with conventional seedbeds.

Emergence and growth of sugar beet on medium and heavy-textured soils also depend on the primary particles being aggregated into larger secondary crumbs, so giving the soil a 'good' structure. However, it is difficult to measure the effects of changes in structure on yield or indeed to change soil structure without affecting some other soil property. Cooke and Williams[67] describe some of the problems of growing sugar beet on the badly-structured Saxmundham soil and elucidate some of the reasons for the 'bad' structure. Experiments have been made in several countries to investigate the value of soil 'conditioners' designed to improve structure, but none is promising at present. Nor does it seem likely that sugar beet will be grown without ploughing or some form of deep cultivating in the near future.

From the somewhat sparse available evidence, it seems that soil physical conditions greatly influence sugar-beet establishment and yield. Much more attention needs to be given to determining which factors are important and how farming operations or soil additives can optimise conditions for germination, emergence and growth. The grower now depends on the scientist to help him decide which and how much plant nutrients are needed for maximum yield, but on experience to decide the cultivations needed to provide the optimum soil physical conditions. It is possible that greatly increased yields may result from cultivations based on the results of future experiments.

Chapter 11

Time, Form and Method
of Fertiliser Application

TIME OF APPLICATION

Most sugar-beet crops receive dressings of the three major nutrients, nitrogen, phosphorus and potassium, just before the seedbed is prepared. Nitrogen is the most important element in fertilisers on most fields because it is usually in short supply in arable land and is used inefficiently by many crops. Although by harvest sugar beet usually takes up more nitrogen than is given in fertiliser, good timing of the application makes sure it is used efficiently during the early stages of growth when the crop is establishing a leaf cover. As nitrate is leached easily, some nitrogen fertiliser must be applied just before, or after, sowing; in many countries more is given as a top-dressing to the growing crop. About a seventh of British sugar-beet growers, for example, top-dress their crops because they consider this counteracts any check to growth from cold or adverse soil conditions or makes good seedbed fertiliser nitrogen assumed lost by leaching. However, there is little evidence that it does any good. As phosphorus and potassium are relatively immobile in soil, time of application is not as critical and numerous tests have been made to determine whether these two elements could be applied before ploughing, rather than in the seedbed, without loss of yield.

Nitrogen

TOP-DRESSING EXPERIMENTS
Sykes[325] showed that top-dressing with 0·6 cwt/acre N as nitrate of soda had little effect on yield of crops given manure before sowing. Ten experiments in SE Scotland in 1954–56[114] also showed that, although top-dressings of 0·4 to 0·8 cwt/acre N increased yield of tops, they had no effect on yield of sugar. However, Williams and Cooke[375] reported that top-dressings increased yield of sugar beet after a very wet spring at Saxmundham.

174

Two large groups of experiments have been made on farms in Great Britain during the last 15 years comparing various seedbed and top-dressing treatments. Adams[2] made 28 experiments comparing 0·6 and 1·2 cwt/acre N applied in the sugar-beet seedbed or as a top-dressing at the end of June and Table 86 shows the average results. The small rate of nitrogen increased yield of sugar most when applied in the seedbed. When 1·2 cwt/acre N was given in the seedbed, yield of sugar was increased slightly, and giving half this amount

TABLE 86

COMPARISONS BETWEEN SEEDBED AND TOP-DRESSED NITROGEN

(after Adams[2] and Last and Draycott[219])

	No nitrogen	N in seedbed	N in seedbed	N as top-dressing	N split-dressing
		Average of 28 experiments, 1956–58			
		0·6	1·2	0·6	1·2
				(cwt/acre)	
Sugar yield (cwt/acre)	41·7	+6·3	+6·8	+4·8	+7·1
Tops yield (ton/acre)	10·0	+3·4	+5·8	+3·6	+6·4
Sugar percentage	16·0	−0·3	−0·7	−0·3	−0·6
Juice purity (%)	88·1	−0·5	−0·3	−0·5	−1·2
		75	150	75	150
				(kg/ha)	
Sugar yield (t/ha)	5·23	+0·79	+0·85	+0·60	+0·89
Tops yield (t/ha)	25·1	+8·5	+14·5	+9·0	+16·1
		Average of 34 experiments, 1959–62			
Sugar yield					
(cwt/acre)	45·9	+9·3	+10·3	—	+8·1
(t/ha)	5·76	+1·17	+1·29	—	+1·02

in the seedbed and half as a top-dressing increased it no further, but there was an increase in the yield of tops. This split dressing decreased juice purity greatly but affected sugar percentage no more than when 1·2 cwt/acre was applied in the seedbed.

Forty-three experiments between 1959 and 1964[219] showed that top-dressings had no significant effect on root or sugar yield (Table 86) but decreased sugar percentage and juice purity. In every year there was a greater chance of depressing yield by splitting the fertiliser application than by applying the full dressing in the seedbed. No crop on soil containing more than 1·7% organic carbon responded to top-dressing. Rainfall and response were not clearly correlated and a regression combining spring and early summer rainfall and percentage organic carbon accounted for a small part of the variation in yield response to top-dressing.

Sugar-beet crops damaged by ectoparasitic nematodes, commonly known as 'Docking disorder', cause fangy, surface root development. Such crops are unable to use nitrogen leached down the soil profile by intense rain after fertiliser application. The nematodes are most numerous, and the damage most severe on light, sandy soils where the leaching may also be most prevalent. Fifteen experiments between 1967 and 1969 were therefore made to determine if top-dressings were needed for affected crops.[60] Although more than the normal amount of nitrogen fertiliser was needed for maximum yield, giving all of it in the seedbed was as good on average as giving part as a top-dressing.

The evidence from Great Britain suggests that all the nitrogen dressing should be given in the seedbed, and top-dressing not practiced. Experiments in France[55] and Ireland[124] have also shown that top-dressings of nitrogen are not justified for sugar beet.

Last and Tinker[217] investigated the nitrogen status of sugar-beet crops given nitrogen in the seedbed only, or part as a top-dressing. The concentration of NO_3^-–N in the leaves was smaller with split dressing early in the season but larger later, which probably accounted for the decreased juice purity of the roots at harvest. Top-dressings on some fields had little effect on NO_3^-–N concentration of petioles in August, because dry weather made the fertiliser unavailable in the dry surface layer of soil. Where plots were irrigated, a top-dressing of nitrogen caused a large increase in NO_3^-–N concentration and decreased sugar percentage at harvest.

EXPERIMENTS WITH ANHYDROUS AMMONIA

In Holland, van Burg et al.[348] compared anhydrous ammonia at nine to five weeks before sowing with the gas injected three to one weeks before sowing. The early application was unsatisfactory but injection just before sowing gave yields comparable with nitrogen applied as calcium nitrate. Draycott and Holliday[87] and Draycott[95] confirmed these results under English conditions, for ammonia injected on the ploughed land six to four weeks before sowing yielded less sugar than ammonia injected into the seedbed, and the seedbed ammonia gave similar yields to 'Nitro-Chalk'. Delaying the application of ammonia until singling time decreased sugar yield compared with the seedbed dressing, but the crop contained as much nitrogen at harvest as that given the seedbed dressing. A liquid compound fertiliser was also tested, either injected into the seedbed or before singling. Delaying the application decreased yield slightly but not significantly.

Phosphorus and potassium

Most experiments testing times of application of phosphorus and potassium for sugar beet have compared autumn with spring applications. Webber[365] made six experiments on light soils in Yorkshire when the fertiliser was either applied half in the seedbed, half ploughed down, all ploughed down or all in the seedbed. On average, the phosphorus and potassium increased yield by 1·3 ton/acre of roots and 4·3 cwt/acre of sugar (Table 87). Seedbed applica-

TABLE 87

EFFECT OF DIFFERENT TIMES OF APPLICATION OF PHOSPHORUS AND POTASSIUM FERTILISER: MEAN OF SIX EXPERIMENTS
(after Webber[365])

	Root yield (ton/acre)	Sugar yield (cwt/acre)	Root yield (t/ha)	Sugar yield (t/ha)
No P or K	12·4	42·3	31·1	5·31
Half P K ploughed down, half in seedbed	+1·3	+4·8	+3·3	+0·60
All P K ploughed down	+0·9	+3·1	+2·3	+0·39
All P K in seedbed	+1·4	+4·8	+3·5	+0·60

tions gave 0·5 ton/acre more roots and 1·7 cwt/acre more sugar than ploughing all the fertiliser down. Splitting the fertiliser dressing gave average yields little different from those obtained with the whole of the dressing in the seedbed.

In Colorado, USA, Nelson[254] obtained the largest yields of roots and total sugar with fertilisers broadcast and ploughed down in autumn, rather than the same fertiliser applied at planting time. Also, Russell[295] found that sugar beet gave a slightly larger yield from ploughed-down phosphorus and potassium on light soils, but the crops on heavy soils yielded best with seedbed fertiliser.

In Finland, Brummer[45] made 31 experiments comparing applications of 2·27 or 4·54 cwt/acre superphosphate and 2·00 or 4·00 cwt/acre potassium chloride before or after ploughing in autumn or in spring. The largest yields were from sugar beet given fertiliser after ploughing in autumn and the smallest where it was given before ploughing. When the large rates were broadcast just before seedbed preparation, the number of seedlings which emerged was decreased by about 10%. Autumn fertiliser before ploughing gave less yield than seedbed applications, particularly where soils contained less than 25 ppm available phosphorus.

The most comprehensive study of time of application of phosphorus and potassium was made by Adams.[3] Experiments throughout Britain compared the effect of all combinations of 1·0 cwt/acre P_2O_5 as superphosphate and 1·5 cwt/acre K_2O as potassium chloride applied either before ploughing in autumn or immediately before seedbed preparations in spring. On average, the spring application gave slightly more sugar and tops (Table 88) than did

TABLE 88

YIELD OF SUGAR AND TOPS FROM SUGAR BEET GIVEN
PHOSPHORUS AND POTASSIUM FERTILISERS IN AUTUMN
OR SPRING: AVERAGE OF 29 EXPERIMENTS
(after Adams[3])

	No P or K	Ploughed down in autumn			Before seedbed cultivations in spring		
		P	K	P K	P	K	P K
Sugar yield (cwt/acre)	43·2	+0·2	+2·3	+3·0	+0·9	+3·0	+5·2
Tops yield (ton/acre)	12·6	−0·1	+0·1	+0·3	+0·1	+0·5	+0·9
Sugar yield (t/ha)	5·42	+0·03	+0·29	+0·38	+0·11	+0·38	+0·65
Tops yield (t/ha)	31·6	−0·3	+0·3	+0·8	+0·3	+1·3	+2·3

ploughing-down. In one year when a small 'starter' dose was given in spring, there was little difference between autumn and spring application. On some plots part of the nitrogen dressing was ploughed down and the sugar beet yielded at least as much sugar as those from plots given all the nitrogen in spring.

Sodium

Crowther[72] described experiments comparing three times of application of sodium for sugar beet. It was either ploughed down in winter, broadcast in winter after ploughing or broadcast a few weeks before sowing. On average, there were only small differences between the three times of application of the sodium, so it can probably be applied at any convenient time up to a few weeks before sowing. Other experiments have shown that it should not be applied less than two weeks before sowing because seedling emergence may be impaired.

FORMS OF FERTILISER

Nitrogen

During the first half of this century, nitrogen in mineral fertilisers applied for sugar beet and other crops was all in the form of ammonium salts and nitrate. Ammonium sulphate (21 % N) accounted for most of the nitrogen used but imported sodium nitrate (16 % N) was also important for sugar beet.[82]

UREA

Over the last 20 years the use of urea ($CO(NH_2)_2$, 46 % N) as a nitrogen fertiliser has increased and some modern compound fertilisers contain at least a portion of the nitrogen in this form. It is also used in foliar sprays because it is unionised and concentrated which gives solutions with a low osmotic pressure, thus minimising the risk of foliar scorch. Urea hydrolyses in soil to ammonia which subsequently nutrifies and is taken up by plants. In alkaline soils nitrogen may be lost to the atmosphere as ammonia gas, so incorporation in the soil is important. Also urea sometimes contains biuret

TABLE 89

AVERAGE YIELD OF SUGAR AND TOPS IN 28 EXPERIMENTS COMPARING AMMONIUM SULPHATE, CALCIUM NITRATE AND UREA

(after Adams[2])

	No nitrogen	Ammonium sulphate	Calcium nitrate	Urea
Sugar yield (cwt/acre)	41·7	+6·1	+6·5	+6·2
Tops yield (ton/acre)	10·0	+4·6	+5·4	+4·6
Sugar yield (t/ha)	5·23	+0·77	+0·82	+0·78
Tops yield (t/ha)	25·1	+11·6	+13·6	+11·6

$((CONH_2)_2NH)$ as an impurity which is toxic to many plants. For foliar application the acceptable level of the impurity is about 0·3 %, but for broadcast soil applications up to 2 % of biuret does not have any adverse effect.

The use of urea for sugar beet has been reviewed by Tomlinson.[337] Adams compared it with ammonium sulphate and calcium nitrate; all were applied to the sugar-beet seedbed before sowing or as a top-dressing at the end of June. Table 89 shows the average results of 28 experiments. All three forms of nitrogen were equally effective in increasing sugar yield and there was no damage to germination. When urea containing 4·5% biuret was used for top-dressing in one year, no damage was seen. Calcium nitrate produced sugar beet with the largest yield of tops.

Devine and Holmes[77] compared urea (containing less than 1% biuret) with ammonium nitrate (33·5% N). They made 19 experiments, mostly on alkaline soils in eastern England, and detected no difference between the mean increase in yield of sugar or tops. However, in other experiments with barley, potatoes and grass, urea gave smaller mean increases in yield than ammonium nitrate or sulphate. The conclusion from these experiments and of many made abroad is that urea gives about the same increase in yield of sugar beet as ammonium and nitrate fertilisers.

AMMONIA

The practical and economic advantages and disadvantages of anhydrous ammonia (NH_3, 82% N) as a fertiliser in British agriculture were reviewed recently.[232] For the sugar-beet crop the main practical advantage lies in its concentration but as it is a gas at normal temperature and pressure, it must be injected into the soil through tines, which is a considerable disadvantage. The main economic advantage is its cost for, per pound of nitrogen, anhydrous ammonia is the cheapest form available although it shows most financial benefit on crops which need more nitrogen fertiliser than sugar beet.

Jameson[194] compared ammonium sulphate with anhydrous ammonia in one field experiment and found no difference in yield of sugar beet grown with the two forms of nitrogen. Where the gas escaped it scorched the growing plants, but the effects were not serious. On the other hand, Hera et al.[172] found that 0·63 cwt/acre N as ammonia gas or ammonia in aqueous solution was more effective than ammonium nitrate—ammonia increased root yields by up to 20% more than ammonium nitrate. However, in experiments in Belgium, Roussel et al.[292] found that anhydrous ammonia gave the same yield in one year and less in another year than ammonium nitrate. Dutch experiments[348] indicated that ammonia gas gave larger yields than calcium ammonium nitrate, whereas, in England, Draycott and Holliday[87] found that 1·0 cwt/acre N as anhydrous ammonia gave the same yield of sugar-beet roots, sugar and tops as 'Nitro-Chalk' although the amount of nitrogen recovered in the crop at harvest was slightly less from ammonia than from 'Nitro-Chalk'.

Thus reports are conflicting on the value of anhydrous ammonia for sugar beet. Comparisons between forms of nitrogen in terms of their effect on sugar yield are only valid where the application was below the optimum, where a range of dosages were applied or where nitrogen uptake by the crop was measured. Discrepancies are also probably due either to partial loss of gas from the injection slit left in the soil, or to differential rates of nitrification in different edaphic and climatic conditions. It seems safe to assume that yields of sugar beet grown with anhydrous ammonia will not differ greatly from those where nitrate or ammonium salts are used, provided there is no loss of gas or damage to the crop.

SLOW-RELEASE FERTILISERS

Measurements of the amount of nitrogen fertiliser leached on a loamy sand cropped with sugar beet showed that losses, particularly from the plough layer, were large in wet springs.[96] Tests were therefore made to determine whether sugar beet would respond favourably on similar soils to forms of fertiliser which released nitrogen slowly.[60] On some soils, isobutylidene diurea showed marked growth and yield increases over calcium ammonium nitrate, but the advantage was not consistent from field to field.

Phosphorus

Surprisingly few experiments have been made comparing different forms of phosphorus for sugar beet. Lachowski[209] compared powdered with granulated superphosphate, both placed and broadcast, in fertilising sugar beet in Poland. There was little difference between the two forms in yield of tops, sugar or in the quality of the roots. Where the powdered form was used, 0·06 cwt/acre P_2O_5 placed gave the same yield of roots and sugar as 0·24 cwt/acre broadcast. Where both forms were broadcast, the sugar-beet seedlings took up more phosphorus from the granulated superphosphate.

Potassium

Gascho et al.[129] compared the yield and quality of sugar beet grown on three contrasting soil types when manured with four potassium fertilisers—potassium chloride, potassium nitrate, potassium sulphate and potassium magnesium sulphate. Yields and quality were the same with all four potassium fertilisers, but giving 1·78 cwt/acre K_2O decreased the quality of sugar beet compared with 0·89 cwt/acre K_2O, which had little effect. Petioles of sugar beet where potassium chloride had been applied contained most potassium, and with 1·78 cwt/acre K_2O compared with those given 0·89 cwt/acre K_2O.

Magnesium

Comparisons between magnesium as sulphate (kieserite), as carbonate (magnesium limestone) and oxide (calcined magnesite) are discussed on page 103.

METHOD OF APPLICATION

Placement compared with broadcasting

Yields of many crops are less from broadcast fertilisers than from the same amount of fertilisers localised near the seed; there are many reviews of this subject.[247,64] The usual method of broadcasting solid fertiliser for sugar beet has been compared with less-conventional methods of application in many experiments in most sugar-beet growing countries.

CONTACT PLACEMENT
Early experiments in England were made by McMillan and Hanley.[237] They mixed fertiliser and seed and sowed them together. This 'combine-drilling' damaged the seed and decreased the emergence. Lewis[225] found that sowing seed and fertiliser in the same row sometimes gave larger yields than broadcasting, especially on soils deficient in one or more nutrients, but he also found the technique involved risks to germination.

BAND PLACEMENT
Lewis[225] also placed fertilisers in a band below the seed and to the side of the seed, and recommended the bands should be $1\frac{1}{2}$ in to the side and 1 in below the seed. Cooke[61] found that contact with a compound fertiliser (9% N, 7·5% P_2O_5, 4·5% K_2O) damaged seed; there was also some damage when fertiliser bands were 2 in below and 1 in to the side of the seed, but fertiliser 3 in to the side was safe. Yields were reduced by methods of applying fertiliser which decreased plant population. There was little difference between mean yields for all experiments given by broadcast fertiliser and fertiliser placed in safe positions near the seed. Placed fertiliser promoted much more vigorous growth of the tops early in the year but by harvest the advantage had disappeared. This is somewhat unexpected as most cultural treatments which increase vigour of sugar beet early in the season usually improve yield.

In ten further experiments,[62] a phosphorus/potassium fertiliser (16% P_2O_5, 13·4% K_2O) was applied in different ways. There was no damage to germination or plant establishment by fertiliser placed in bands 2 in to the side and 2 in below the level of the seed. There were no significant differences between the yields of sugar given by placed and broadcast fertiliser. Similar yields were given by broadcast applications applied in early spring and worked into the seedbed, and by dressings on the seedbed which were worked in shallowly. In most of the experiments, placement again gave more vigorous growth during late spring and early summer than broadcasting the fertiliser, but by harvest the superiority had disappeared. Shotten[311] made 16 similar experiments in Norfolk to compare a compound fertiliser (average analysis: 10% N; 10% P_2O_5; 15% K_2O) broadcast with the same fertiliser placed by commercial drills. Placement of the fertiliser produced larger yields than broadcasting on 11 fields, gave the same yields on three fields and was inferior to broadcasting on two fields where the broadcast fertiliser was deeply incorporated in the soil. The average increase in yield from placement was 0·43 ton/ acre roots or 1·7 cwt/acre sugar. Germination was not impaired by placement and final plant populations were larger when the fertiliser was placed.

Prummel[281] summarising the results of Dutch experiments on placement of nitrogen, phosphorus and potassium showed that it was of greater benefit for cereals and pulses than for sugar beet. For sugar beet he found that if the effectiveness of broadcast fertiliser was taken as 1·00, the relative effectiveness of band placement of nitrogen was 1·20, phosphorus 1·20 and potassium 1·00. Nitrogen damaged the seed if placed nearer than 2 in; also, placed nitrogen prolonged the period of growth of sugar beet compared with broadcasting.

The annual and residual effects of fertiliser placement were investigated in a long-term experiment by Hanley and Ridgman.[157] Sugar beet was given a relatively small dressing of fertiliser. The effect of placement was variable and sometimes reduced yields. The greatest reduction was in 1957, a year when spring rainfall was unusually small (0·80 in during April and May) and placement had a harmful effect on plant establishment. Analysis of soil showed that placing a small quantity of fertiliser decreased soil reserves of phosphorus and potassium compared with broadcasting twice as much fertiliser.

Summarising the results of experiments made on sugar beet comparing broadcasting with placement of fertiliser, on average, there was little advantage in placement. Damage to germination, with decreased plant population and consequent loss of yield, were a feature of many of the experiments where fertiliser was placed too near to the seed, particularly in dry springs. The 'safe' distance seemed to be at least 3 in to the side and 2 in below the level of the seed. Where damage to establishment was avoided, small increases in yield of the order of 2 cwt/acre sugar were obtained. Now that most crops are sown with precision drills, it is important not to disturb the soil at depth near the line of seed sowing; also, if wet soil is brought near the surface, seeds are not delivered accurately. Speed of sowing is all-important, so that the crop is sown as early as conditions allow; thus there appears to be little prospect for placement.

Liquid fertilisers

Recent advances in technology offer alternatives to solid fertilisers; compounds containing nitrogen, phosphorus and potassium are obtainable as concentrated aqueous solutions, and nitrogen can be bought cheaply as anhydrous ammonia. These materials have been used for some time in other countries, particularly the United States of America, and are now readily available in Great Britain.

Liquids can be sprayed on the soil surface, or injected accurately into the soil through tines in a band at the required distance and depth below the seed or young plant.[81] Few experiments have been made testing liquid fertilisers on the sugar-beet crop, but with other crops it has been shown that yields from sprayed liquids and from broadcast solids are the same provided the crop is not damaged and that both fertilisers are applied at the same time.

Devine[76] compared solid fertiliser with a liquid fertiliser of similar composition and nutrient ratio for sugar beet. The solutions containing ammonium nitrate, ammonium phosphate and urea were applied by means of a knapsack sprayer adjusted to give large droplets. There were no significant differences in yield of the sugar beet between liquids and solids. Draycott and Holliday[87] reported similar findings with a liquid compound fertiliser (average analysis: 17% N; 8% P_2O_5; 15% K_2O) compared with a solid granular compound of the same analysis. The liquid was sprayed on the seedbed and the compound broadcast by hand. Both were worked into the soil and both gave the same yield of sugar from the sugar beet. They also compared the sprayed liquid with the same liquid injected to give a band of fertiliser, either 2 in to the side and 2 in below the seed, or 2 in to the side and 6 in below the seed. Yields of sugar from the sugar beet grown with fertiliser applied by the three methods were not significantly different, but the deep placement gave a small consistent increase in yield over shallow placement, probably due to increased availability of the deep-placed fertiliser as it was in soil which did not dry as readily as the surface soil. This was substantiated by a greenhouse pot experiment with sugar beet.

Foliar application of nitrogen

Desprez[78] compared the effect of urea solution sprayed on the leaves of sugar beet in France at 0·18 cwt/acre N with the same amount of nitrogen as solid sodium nitrate broadcast. Both were applied three weeks after sowing in addition to the conventional dressing of 1·20 cwt/acre N applied in the seedbed, and late sprays of urea were included in later experiments. The results indicated that provided the urea was applied at least three months before harvest it did not decrease sugar percentage. Unfortunately it increased root yield little. Whereas urea was assimilated by the plants almost at once, nitrogen from the broadcast sodium nitrate was taken up later and consequently decreased sugar percentage.

Thorne and Watson[328] found that when leaves of sugar-beet crops were sprayed in September and October with ammonium nitrate or urea solution, 70% of the nitrogen was recovered in the plants compared with 40% from soil applications. Spraying slightly increased the dry matter yield of the tops, but not of the roots. It decreased the sugar percentage of the roots by 1%. Kozera and Lachowski[205] in Poland sprayed nitrogen, phosphorus and potassium solutions on the foliage of sugar beet several weeks before harvesting. The solutions increased yield of tops slightly but did not affect yield of roots or sugar percentage. There seems little scope for increasing sugar yield by major nutrient sprays in late summer or autumn.

Conclusions

There is much evidence that sugar beet must be well-supplied with nitrogen during the early stages of growth; in Great Britain this period is from emergence, which is usually in April, until complete leaf cover in June. However, there is very little evidence that giving nitrogen fertiliser after June ever increases sugar yield, and it always decreases sugar percentage and juice purity. Nitrogen fertiliser leaches rapidly during winter and spring in the climate of NW Europe, particularly when applied in nitrate form, so some nitrogen fertiliser needs to be applied for sugar beet just before, at or just after sowing. If heavy rain follows, an immediate supplemental dressing to make good leaching losses may be useful, but the overwhelming evidence from many experiments is that top-dressings rarely increase yield. They give slight benefit where some factor, such as poor seedbed or soil conditions or where the root system is damaged, prevents normal growth and uptake of nitrogen. With these possible exceptions, the whole of the nitrogen fertiliser should be given during the period from two weeks before to two weeks after sowing.

In countries where the crop is sown in autumn and harvested the following year, nitrogen fertiliser practice must depend on the amount and distribution of rainfall. In Mediterranean climates, where much sugar beet is grown in this way, some nitrogen fertiliser is given in the seedbed to ensure good autumn growth, and some after the winter rain when the crop begins to grow rapidly again. In similar climates, nitrogen is often supplied in irrigation water, but in all countries the supply of nitrogen fertiliser should cease two to three months before harvest.

The timing of applications of phosphorus and potassium fertilisers is much less critical than for nitrogen, for most soils where sugar beet is grown usually contain large residues of phosphorus and potassium from fertiliser given previously. As these two elements are not normally leached, the 'new' fertiliser merely supplements 'old' fertiliser. Loss through permanent fixation in near-neutral arable soils is also of doubtful importance. Thus applications of these two elements can be made when most convenient for the grower; provided that the soil contains a moderate supply of available phosphorus and potassium (*e.g.* more than 10 ppm sodium bicarbonate-extractable P and 100 ppm ammonium nitrate exchangeable K), fertiliser can be applied at any time between harvesting the previous crop and sowing the sugar beet. The best time for wheeled spreaders is after harvest prior to ploughing. In Great Britain this task is being undertaken increasingly by contractors, and in some countries application is now from aircraft. In this way there is no compaction of the ploughed land and no delay in sowing.

Many comparisons have been made between different forms of fertiliser for sugar beet, particularly between forms of nitrogen. The conclusion is that the common forms used in fertiliser (*e.g.* ammonium sulphate, ammonium nitrate, urea, anhydrous ammonia) are equally satisfactory. Some slow-release forms of nitrogen (*e.g.* isobutylidene diurea and 'Nitro-form') have given promising results on soils prone to leaching but more experiments are needed before they can be recommended for commercial use.

Few comparisons have been made between different forms of phosphorus. Results of experiments with other crops lead to the conclusion that the element needs to be soluble in water or citric acid. Rock phosphates only supply phosphorus to crops slowly, particularly under the neutral or alkaline conditions of sugar-beet soils. The softer 'Gafsa' rock phosphates may be worthy of further investigation now that the phosphorus fertiliser practice on many farms merely aims to replenish offtake. Potassium is applied in the form of chloride either in crude salts such as kainit, or in compound fertilisers. Sodium is also largely applied as chloride, but sodium (Chilean) nitrate is also a convenient form. There is little evidence that giving potassium or sodium in a form other than the chloride would be beneficial for sugar beet.

Reports on experiments comparing broadcasting with placement of fertiliser provide little support for placement. As broadcasting is the quicker, simpler and cheaper method it seems unlikely that fertiliser will ever be placed for sugar beet. Likewise there are no results to indicate that foliar application of nutrients has any benefit over applying them to the soil—very often the latter method is more satisfactory.

Chapter 12

Irrigation, Plant Density, Pests and Diseases

Irrigation increases yield directly by supplying water needed for transpiration and indirectly by improving availability of nutrients, so the quantity of fertiliser required by the sugar-beet crop may be influenced both by the increased yield of the irrigated crop, and by changes in availability of nutrients present in the soil. Similarly, changing the plant density greatly affects the growth and yield of individual plants and the depth and distribution of their roots in the soil. This, too, affects the uptake of nutrients and the amount of fertiliser needed for maximum yield. Pests and diseases slow the growth of the crop and more fertiliser is often needed by affected crops than by healthy ones.

IRRIGATION

IN GREAT BRITAIN

In four experiments on free-draining sandy loams Garner[126] and Penman[272] grew the crop with 0·4 or 0·8 cwt/acre N. The additional nitrogen increased yield slightly without irrigation but decreased it with irrigation. Price and Harvey[279,280] investigated the effects of irrigation on response to nitrogen (0, 0·5 and 1·0 cwt/acre), phosphorus (0·6 and 1·2 cwt/acre P_2O_5) and potassium (0·6 and 1·2 cwt/acre K_2O) in the years 1955 to 1959 on sandy soil at Gleadthorpe, Nottinghamshire. In the first three experiments 1 in water was applied when the deficit reached 2 in and for the last two years of the investigation the soil was watered at drilling and to field capacity whenever the deficit reached 1 in. Three out of the five years were wet and there was little gain from irrigation but in 1955 and 1959, which were exceptionally dry seasons, the responses to watering were so large that the mean return over the five years was profitable.

Table 90 shows the effect of irrigation in response to nitrogen fertiliser in the two dry seasons. Irrigation without nitrogen greatly increased root and sugar yields but decreased sugar percentage. Nitrogen increased root yield slightly and irrigation did not improve the response; however, nitrogen decreased sugar percentage less with water than without. 0·5 cwt/acre N was the best dressing for sugar yield both with and without irrigation, and there was no evidence that irrigation affected the amount of phosphorus or potassium required for maximum yield.

TABLE 90

EFFECT OF IRRIGATION ON RESPONSE TO NITROGEN BY SUGAR BEET ON A SANDY SOIL: MEAN RESULTS OF THE DRY SEASONS, 1955 AND 1959
(after Price and Harvey[280])

N dressing	Root yield		Sugar percentage		Sugar yield	
	Without water	With water	Without water	With water	Without water	With water
(cwt/acre)	(ton/acre)				(cwt/acre)	
0	4·8	19·6	19·2	18·1	18·4	71·8
0·5	5·0	20·8	18·5	18·3	18·9	76·2
1·0	5·2	20·4	18·0	18·0	18·3	74·4
(kg/ha)	(t/ha)				(t/ha)	
0	12·1	49·2	19·2	18·1	2·31	9·01
63	12·6	52·2	18·5	18·3	2·37	9·56
126	13·1	51·2	18·0	18·0	2·30	9·34

Penman[273] measured response to irrigation and nitrogen by sugar beet in each of the years 1951–59 at Woburn. No irrigation, full irrigation to near field capacity and early and late irrigation were tested and the plots were divided for nitrogen dressing—0·4 and 0·8 cwt/acre nitrogen in 1951–56 and 0·6 and 1·2 cwt/acre 1957–59. There was some evidence that the crop gave most sugar with the larger dressing *without* irrigation and with the smaller dressing *with* irrigation. In a further three years of experiments at Woburn, 0·75 and 1·50 cwt/acre N was applied[274] and irrigation had little effect on response to nitrogen.

In an experiment on the Reading University farm, Harris[160] found that sugar beet only responded to nitrogen without irrigation. Irrigation made the plants look greener and increased their nitrogen concentration, which probably explains the lack of response to nitrogen by the irrigated crop. In experiments at Broom's Barn[97] (*see* page 193), irrigation had little effect on the amount of nitrogen needed for maximum yield (Table 91) even where crops responded greatly to nitrogen and in dry summers.

TABLE 91

EFFECT OF IRRIGATION ON NITROGEN REQUIREMENT
BROOM'S BARN: MEAN OF 5 YEARS AND 4 PLANT DENSITIES
(after Draycott and Webb[97])

N dressing		Without	Irrigation With Without		With
(cwt/acre)	(kg/ha)	(cwt/acre)	Sugar yield	(t/ha)	
0	0	51·4	50·5	6·45	6·34
0·6	75	62·0	62·6	7·78	7·86
1·2	150	63·2	66·5	7·93	8·35
1·8	225	60·8	64·8	7·63	8·13

IN THE USA

As in England, most experiments have investigated interactions
between nitrogen and irrigation but a few have included additional
plots with other major nutrients. In contrast to the English experiments, most of the American experiments have been made where the
sugar-beet crop is irrigated in commercial practice. Consequently, in
most experiments, all plots have been given some irrigation and the
treatments have either been additions to this or have tested different
methods of application of water. The majority of the results substantiate the finding of the English experiments that irrigation does
not affect nitrogen fertiliser requirement greatly.

Working in Utah, Haddock[146] grew sugar beet on a deep well-
drained loam and compared sprinkler and furrow irrigation. Plots
were subdivided and given an additional 0·7 cwt/acre N and 0·4
cwt/acre P_2O_5. Nitrogen did not increase yield but the phosphorus
fertiliser gave a large increase, particularly with sprinkler irrigation,
suggesting that the availability of the phosphorus was increased by
irrigation. Henderson et al.[168] substantiated this by comparing sub-
irrigation with sub-irrigation plus sprinkler irrigation in California.
With sub-irrigation only, the surface soil was relatively dry for the
major part of the growing season, but the crop was well-supplied
with water for transpiration. Sprinkler irrigation nevertheless
increased nitrogen and phosphorus concentrations in the sugar-beet
petioles and increased sugar yield. They concluded that the response
to sprinkler irrigation was partially due to improved water supply
and partly to more favourable conditions for phosphorus and nitrogen uptake. On soil where nitrogen greatly increased yield, Woolley
and Bennett[379] investigated the amount of nitrogen needed for

maximum yield when grown with four amounts of irrigation. Compared with giving none, 0·7 cwt/acre N increased root and sugar yield with all the moisture treatments. However, giving 2·23 cwt/acre N increased yield further only with the largest amount of water and decreased yield on plots given little water.

Robins et al.[289] tested the effect of excessive irrigation on the nitrogen requirement of sugar beet in the Columbia Basin, Washington. Early excess water, i.e. more than was needed to bring the soil to field capacity, decreased yield and increased the amount of nitrogen fertiliser needed by the crop; they concluded that this was mainly due to leaching, for the concentration of plant nitrate was decreased and sugar percentage increased by the excess of water. When excess water was given in autumn, yields were unaffected but the nitrogen status of the plants was decreased.

Loomis and Worker[229] in California showed that moisture stress and nitrogen deficiency affected sugar-beet growth in similar ways. Both decreased vegetative growth and increased the sugar percentage in the roots; purity and sugar yield were increased by decreasing the nitrogen dressing, though not by increasing moisture stress. The effects of the two factors were independent and additive. In California where the crop is irrigated, sugar beet is often allowed to wilt for a few weeks before harvest, supposedly to concentrate the sugar in the roots. Loomis and Worker found that the practice did not increase sugar yield or quality in their experiments, although in commerce there might still be some advantage through lower haulage costs.

PLANT DENSITY

There have been many investigations into the effects of changes in plant density of sugar beet but surprisingly few have tested the effect of different quantities of fertiliser on the plant density/yield relationship.

IN GREAT BRITAIN
Harris[159,160] described two investigations at the Reading University farm where singling and harvesting methods and nitrogen treatments (0–1·28 cwt/acre) were tested. Machine-singling gave populations between 32 and 66 thousand/acre and hand-singling 26 to 30 thousand/acre. Nitrogen had no effect on yield and it did not interact with any other treatment. Machine-harvesting was particularly ineffective in harvesting the machine-thinned crop for there were

many small roots. However, increasing the nitrogen dressing did not improve the performance of the harvester compared with hand-harvesting.

In the second report,[160] the main object was to investigate the interaction between plant density and irrigation (*see* page 193), but nitrogen was also tested. Harris showed that large plant stands did not need nitrogen fertiliser, which was surprising for colour scores indicated that increasing plant density decreased the nitrogen status of the crop. Analysis of plant samples at Broom's Barn has confirmed that the crop contains a smaller concentration when grown in dense stands than when the plants are widely spaced. This indicates that roots in dense stands compete for nitrogen but sugar yield is not increased by giving more; this is because additional nitrogen increases yield of tops but not the yield of roots.

IN THE USA

Some experiments have shown that increasing plant density has *no* effect on nitrogen requirement, some that it *decreases* requirement

TABLE 92

MEAN SUGAR YIELDS IN TWO EXPERIMENTS IN KANSAS, USA FROM THREE PLANT DENSITIES GIVEN THREE AMOUNTS OF NITROGEN FERTILISER
(after Herron *et al.*[174])

Plant density		N dressing					
			(cwt/acre)			(kg/ha)	
(1 000/acre)	(1 000/ha)	0	0·63	1·16	0	79	146
			Sugar yield				
			(cwt/acre)			(t/ha)	
18	7·3	85·3	85·2	84·4	10·7	10·7	10·6
24	9·7	86·2	85·9	81·3	10·8	10·8	10·2
36	15·0	84·8	87·2	84·6	10·6	10·9	10·6

and others that it *increases* nitrogen requirement. For example, Herron *et al.*[174] in Kansas grew sugar beet on 22 in rows and singled the brairds to give 8, 12 and 16 in within row spacing, resulting in populations of 36, 24 and 18 thousand plants/acre respectively. The 12 in spacing gave the largest root yield and 8 in spacing the smallest in both years (Table 92). All the spacings were tested with 0, 0·7 and 1·4 cwt/acre N in the first year and 0, 0·54 and 1·08 cwt/acre in the second year. Each increment of nitrogen increased root yields, but response was not affected by plant spacing. Nelson[253] in Arizona reported a significant negative interaction between plant spacing of

5, 10 and 15 in on an average row width of 20 in (60, 30 and 15 thousand plants/acre) and nitrogen fertiliser (*see* below):

N dressing (cwt/acre)	Plant spacing (in)		
	5	10	15
	Root yield (ton/acre)		
0·63	19·6	20·0	18·7
1·56	21·6	23·8	22·6
2·23	20·9	23·5	23·8

Nelson expected the narrow-spaced plants would need most nitrogen but they responded least; since the crop received adequate moisture, he suggested that a shortage of light was limiting yields of the close-spaced plants. Quite different results were recorded by Haddock and Kelley[142] in Utah who grew sugar beet in rows 16, 20 and 24 in apart with a constant spacing of 12 in (33, 26 and 19 thousand plants/acre respectively). Various fertilisers were tested, including 0, 0·71 and 1·42 cwt/acre N. Nitrogen increased sugar yield significantly with all plant spacings and there was a large positive interaction between increased plant density and increased nitrogen application. The authors concluded that to obtain maximum return from large dressings of nitrogen it is necessary to have enough plants to utilise it fully.

IRRIGATION AND PLANT DENSITY

IN GREAT BRITAIN
Draycott and Webb[97] investigated the interactions between nitrogen fertiliser, plant population and irrigation on sugar beet at Broom's Barn for five years. On average, nitrogen and plant population influenced yields greatly but irrigation relatively little. Although total dry matter was greatest when the largest plant population was given the largest dressing of nitrogen and irrigation, the proportion of dry matter partitioned to the roots was decreased by all three factors. Irrigating the smallest plant population gave no increase in yield without nitrogen and only a small increase with 0·6 cwt/acre. However, as the nitrogen dressing was increased, response to irrigation increased, but irrigating the largest plant population increased yield equally with all amounts of nitrogen. The nutrient concentration and uptake by the crop were also greatly affected,[98] for nitrogen fertiliser and irrigation increased uptake of nitrogen by the crop but increasing the plant population had little effect on uptake and decreased the concentration of nitrogen.

It is concluded that competition for light in the dense stands was the main limiting factor to increased yield of sugar; giving more nitrogen and water failed to increase sugar yield and tests of other major nutrients showed that they were not limiting yield. Competition for light appeared to stimulate the growth of leaves and at harvest the proportion of dry matter partitioned to the roots was least with the largest plant density. Irrigated sugar beet in England give maximum yield of sugar when grown in plant densities of 25–40 000 plants/acre with 0·6–1·2 cwt/acre N.

In the USA

Haddock and Kelley[142] investigated the effects of water and nitrogen applications for sugar beet grown at three plant densities in Utah. With adequate water, the closest spacing (12 × 16 in) produced the largest yield of sugar and needed the largest amount of nitrogen (1·42 cwt/acre) for maximum yield; giving less water decreased yield, but watering did not affect the amount of nitrogen needed. Haddock[143] reported results of soil moisture measurements in the same experiments which indicated that early watering was much more important than late watering with all plant stands.

Water requirements of sugar beet

It is generally accepted that sugar beet, like most crops, uses water at a rate proportional to evaporation from an open water surface. Penman[271] first showed that the amount of water used by a crop which completely covers the ground with leaves and is not short of water, transpires water at a rate proportional to evaporation calculated from meteorological measurements. Vilain and Avronsart[351] made measurements of the amount of water used by sugar-beet crops grown in France and compared them with the theoretical requirement estimated from meteorological measurements. As the crop does not completely cover the soil surface at the beginning of the season, the amount of water used is less than the theoretical amount, and if the crop is short of water at any period during its growth, then the amount of water transpired may be less than the theoretical amount.

However, Vilain and Avronsart showed that sugar beet grown in a heavy clay soil was able to extract a large amount of the required water from the soil. With normal rainfall uniformly distributed throughout the year, the crop obtained sufficient water from the soil to produce a yield of nearly 72 cwt/acre sugar, and measurements made in the soil profile at various depths from the surface down to

63 in showed that the roots extended rapidly down to the full depth of measurement and extracted water from all the horizons.

Similar measurements were made on the British crop, where winter rainfall usually exceeds that required to return the soil to field capacity.[92,99] The soil is usually at field capacity at the end of March, and during April, before the sugar-beet crop establishes a leaf cover, water is only lost by evaporation. Until a deficit of 1 in is reached, the loss is approximately equal to the potential transpiration calculated from meteorological data. If the soil is not disturbed by cultivations, evaporation thereafter is only a small fraction of potential transpiration. During May and June (until 100% leaf cover) water loss from the soil profile is partly by evaporation from bare soil but increasingly by transpiration through the crop. When a complete leaf cover has formed, the soil moisture deficit increases at the same rate as the potential transpiration, less rainfall, until water in the profile limits loss by transpiration.

In two out of three years the water-use by sugar beet closely paralleled the calculated evapotranspiration after the establishment of a complete leaf cover on plots which were not irrigated. In one year when the summer was very dry, the crop could not obtain sufficient water for transpiration and the water-use was less than calculated from meteorological data. In all three years moisture was lost from the surface 8 in mainly by evaporation from bare soil during the period up to singling. During the latter part of May and in early June, sugar beet withdrew moisture down to 32 in. By early July, water was extracted down to 40 in, and in the two drier years there was evidence that roots took out up to 2 in moisture from below 40 in. Results during the latter part of the growing season suggested that with deficits under sugar beet of less than 6 in the root system sustained a water requirement of the crop from soil reserves, except when the deficit increased rapidly in one year. This no doubt explains the small response to irrigation by sugar beet on most soils in England.

When the soil moisture deficit increased rapidly early during the season the crop extracted water from the soil by exhausting the available water from progressively deeper horizons, whereas when the deficit increased rapidly late during the season, water was still being extracted from all horizons until harvest. Both decreasing the plant population and irrigating decreased the amount of water used from depth in the profile every year.

The total amount of water used (evaporation + transpiration) on average from soil reserves and rainfall was 12 in by a small plant population and 14 in by a large population. When irrigated, the consumption increased to 14 in and 16 in respectively. The difference between uses between populations was almost entirely from the difference from leaf cover early in the season. The water consumption in a very wet summer was only two-thirds of that in relatively dry sunny summers. On average, the crops produced 6 ton/acre of total dry matter and transpired 14 in of water. Thus the dry matter production per unit of water transpired was 8·5 cwt/acre/in; comparable sugar production was 5 cwt/acre/in. Where irrigation was given, it increased sugar yield by about 2·5 cwt/acre/in on average.

PESTS AND DISEASES

Where shortage of a nutrient decreases vigour and yield, such crops are often more susceptible to attack by pests and diseases. Fertilisers may also affect the soil-borne pathogens and the crop's tolerance to pests and diseases. Where pathogens damage the root system, nutrient uptake is affected and fertiliser requirement increased. Examples of some of these interactions between pests and diseases, and the nutrition of the crop are discussed below.

Effect of fertilisers on inoculum in soil

Sclerotium root rot is a serious disease of sugar beet in some countries, being caused by the fungus *Sclerotium rolfsii*, the white mycelium of which covers the surface of affected roots. Leach and Davey[220] in California showed that applications of nitrogenous fertilisers decreased the percentage of plants affected by this fungus. Ammonium sulphate, anhydrous ammonia and calcium nitrate were equally effective when applied in equivalent amounts of nitrogen. The percentage of roots affected was halved by 1·0 cwt/acre N and the fertiliser greatly increased yield. Laboratory trials showed that ammonium solutions were toxic to the mycelium and sclerotia but failed to explain the effect of the fertilisers in the field, for calcium nitrate was not toxic in the laboratory tests. The partial control in the field may have been due to changes in the metabolism of the causal fungus which decreased its growth, to increased resistance by the sugar beet or to changes in the balance of micro-organisms in the soil.

A somewhat similar widespread disease known as 'violet root' rot is caused by the soil-inhabiting fungus *Helicobasidium purpureum*. Hull and Wilson[185] showed that 0·8 cwt/acre N as ammonium sulphate decreased the number of roots infected by 25%; phosphorus and potassium fertilisers had no effect, but organic manure and sodium were beneficial. Crop rotation also affected the incidence of the disease for carrots, sugar beet, mangolds, other root crops and some species of weed act as hosts to the fungus.

Effect of fertilisers on tolerance and symptom expression

Hull and Watson[186] made experiments to determine the losses from beet yellows virus when the sugar beet was grown with different amounts of fertilisers and farmyard manure. The fertilisers increased the root and sugar yields of both infected and healthy plants but the losses caused by infection increased proportionally as the mean yields increased and there were no significant interactions on average. The fertilisers had little effect on the symptoms of the disease or on rate of spread of infection; nitrogen, phosphorus and potassium occasionally increased the rate of spread and agricultural salt consistently decreased it. Adams[8] in a large series of nitrogen trials recorded the proportion of plants with virus yellows at the end of August. There was little relation between yellows incidence and response to nitrogen, indicating that the disease was not a major factor affecting response. Similarly, in experiments described by Tinker[329] the mean response to nitrogen fertiliser was unaffected by different amounts of virus infection.

The effect of nitrogen fertiliser on the spread of yellowing viruses—beet mild yellowing virus (BMYV) or beet yellowing virus (BYV)—and the appearance of symptoms on the leaves of sugar beet have been studied by Heathcote[164, 165] in experiments testing 0–1·80 cwt/acre N, which had a large effect on the leaf area index and the height of the plants. Nitrogen tended to increase the number of aphids (*Myzus persicae* Sulz. and *Aphis fabae* Scop.) which are vectors of the viruses, but this effect was not entirely consistent. He suggested that the increased numbers may have been due to the increased plant height or leaf area, or to increased fecundity of the aphids caused by the increased concentration of nitrogen in the plant sap. However, all plants were later sprayed with aphicide (common practice on commercial crops in England), so most of the virus-spread was by winged aphids. Hull[188] reported that large dressings of nitrogen fertiliser masked symptoms of beet mild yellowing virus but Blencowe and Tinsley[25] found that the percentage of plants showing symptoms was greater in plants with more nitrogen fertiliser, but it was not clear whether they were observing BMYV or BYV. In recent experiments with trace elements, Russell[296] found that aphids settled more readily on plants given lithium, zinc and nickel but less readily on those treated with boron. Viruliferous aphids transmitted BYV more efficiently to plants with lithium and boron than those with copper, zinc and tin.

Effects of fertilisers on root development

Young,[384] on infertile soils in Ohio, found that Black root incidence decreased as soil organic matter content was increased. Phosphorus fertiliser was very effective in controlling the disease, particularly when placed in a band near the seed. Nitrogen and potassium fertilisers helped to decrease the damage caused by the disease but neither was effective when phosphorus was also present.

The stunting of sugar-beet plants in patches on sandy soils, when caused by *Longidorus* and/or *Trichodorus* feeding on the seedling roots, is known as 'Docking disorder'.[367] Some of the yield loss is because damaged roots do not absorb enough nutrients, and in early experiments nitrogen in various forms and organic manures sometimes increased yield greatly but magnesium had no effect. Dunning and Winder[107] in recent experiments found that only nitrogen of several fertilisers increased yields when placed in the root zone of stunted plants. In an experiment in Yorkshire, where the crop was given 1·2 cwt/acre N in the seedbed, an additional 1 cwt/acre N doubled sugar yield.

Cooke and Draycott[60] tested 0, 0·66, 1·32 and 1·98 cwt/acre N applied to the seedbed as 'Nitro-Chalk' on 15 fields in the presence and absence of fumigation with 'D-D' (dichloropropane-dichloropropene mixture). Without fumigant, the crop responded up to the largest amount of nitrogen tested, for the root systems of many plants were severely damaged. Yield was greatly increased by fumigation and the crop needed only 0·66 cwt/acre N for maximum yield; on average of 15 experiments, the sugar yields were:

| | N applied (cwt/acre) | | | |
	0	0·66	1·32	1·98
	Sugar yield (cwt/acre)			
Without fumigant	33·7	41·9	44·8	46·2
With fumigant	49·4	58·3	58·2	58·3

In four of the 15 experiments, nitrogen supplied in a slow-release form (isobutylidene diurea) gave much greater yields than the same amount of nitrogen as 'Nitro-Chalk', suggesting that nitrogen fertiliser given in the seedbed in soluble forms may be leached. However, fumigation not only improved the health of roots, and so enabled them to use nitrogen more efficiently, but also increased the amount of available nitrogen in the soil and decreased the amount lost by leaching.

Sugar beet may be defoliated at any time during the growing season by hail or pests, by soil particles during wind erosion in spring, by weed cutting in summer or by topping before harvesting. Dunning and Winder[109] artificially defoliated field-grown plants completely each month from May to September on plots with and without 150 kg/ha N fertiliser. At harvest in mid-November, minimum root weights followed defoliation in July/August and later defoliation gave minimum sugar percentages. With nitrogen, sugar yields were smallest after August defoliation but in the absence of nitrogen they were smallest after July defoliation. Up to 40% of the sugar yield was lost by defoliation in these two months but yields, and recovery from defoliation, were greater with nitrogen than without.

Conclusions

In NW Europe large yields of sugar beet are obtained without irrigation because rainfall during the growing season plus soil reserves of water are nearly sufficient to satisfy the needs of the crop in most years. The deep and spreading root system of healthy sugar beet allows 6 in or more of water to be taken from soil reserves and this is replenished by winter rainfall. However, supplementary irrigation increases sugar yield in dry summers, particularly on sandy soils or where the root system is damaged by pests or diseases or where its development is impeded by poor soil physical conditions. Irrigation, even if it greatly increases yield, does not appear to affect the amount of fertiliser needed for maximum sugar yield. That the fertiliser requirement is unchanged is substantiated by experiments in drier climates, where changing the amount of water given to irrigated crops has little effect on the amount of fertiliser needed for maximum sugar yield. Although water shortage and nitrogen deficiency affect sugar-beet growth in similar ways, the increases in sugar yield from water or nitrogen fertiliser are independent and additive.

As the number of plants per unit area is increased the concentration of nitrogen in the dry matter is decreased; this effect can be observed in the field because leaves of plants in dense stands are light green in colour compared with the dark green foliage of widely-spaced plants. This indicates that plants in dense stands compete for nitrogen, and it has been assumed that more nitrogen fertiliser is needed if dense stands are to give maximum sugar yield. Experiments have shown that this is not so, for giving more nitrogen fertiliser increases yield of tops without increasing the yield of roots or sugar. As with irrigation, changing the plant density appears to have little effect on the amount of nitrogen fertiliser needed for maximum sugar yield. It is concluded that competition for light in stands of more than 30 thousand plants/acre is the main limiting factor to increased sugar yield. Giving more nitrogen fertiliser and/or water does not help— they simply stimulate the growth of tops without any commensurate increase in sugar yield.

Chapter 13

Sugar Beet Quality

With most crops it is difficult to define quality, for subjective criteria such as texture, taste, shape and colour are involved. Fortunately with sugar beet, most of the important aspects of quality can be defined and measured easily. The two most important features are evaluated chemically—the sugar percentage and the purity of the root juice.[52] Other physical characteristics important in harvesting and processing, such as the root shape and the ease of slicing,[284] are more difficult to define but are affected little by fertilisers.

Determination of sugar percentage

The amount of sucrose in the root is normally determined polarimetrically on a lead acetate extract of fresh macerated root ('brei') by the method first used by Sachs and described by Le Docte.[221] The 'sugar percentage' so measured is usually in the range 10–20%. Fertilisers have considerable effects on sugar percentage—some decrease it but some are beneficial. Where a fertiliser decreases sugar percentage but increases root yield it is important to know the 'break-even' point where the increase in root yield is equated with decrease in sugar percentage, *i.e.* maximum sugar production by the crop.

The sugar percentage of fresh roots is, of course, largely determined by the amount of water in the roots, and climatic conditions before harvest, therefore, have a marked effect. Fertilisers also affect the water content of the roots and thereby the sugar percentage. The concentration of sugar in the roots expressed as a percentage of the root dry matter removes the effects of changes in water content and few reports have shown that some fertilisers also slightly affect the amount of sugar in the dry matter.

Determination of juice purity

Juice purity is determined in the laboratory by a combination of refractometric and polarimetric measurements on juice expressed from fresh brei.[52] Another method of assessment has been described by Carruthers et al.[53] Sodium, potassium and α-amino nitrogen concentrations in the lead acetate extract (prepared for the determination of sugar percentage) are determined and the values used in the regression line: Juice Purity = 97·0 − 0·000 8 (2·5 K + 3·5 Na + 10 α-amino N), K, Na and N being expressed as mg per 100 g sugar. Carruthers et al.[53] found a close relationship between the two assessments of juice purity ($r = 0·86$).

Effect of nitrogen fertiliser

IN GREAT BRITAIN
Draycott and Cooke[80] reporting over 400 field experiments 1934–49 showed that nitrogen fertiliser considerably diminished sugar percentage of sugar beet; compared with phosphorus and potassium fertilisers the effects were:

$$N: -0·38$$
$$P: +0·02$$
$$K: +0·24$$

In further experiments 0·6, 1·2 and 1·8 cwt/acre N was tested with and without farmyard manure and Table 93 shows that both nitrogen and farmyard manure depressed sugar percentage and juice purity, but there was no interaction between them.

TABLE 93   Table 13·1

EFFECT OF NITROGEN FERTILISER AND FARMYARD MANURE ON SUGAR PERCENTAGE AND JUICE PURITY
(after Draycott and Cooke[80])

	N dressing (cwt/acre)		
	0·60	1·20	1·80
	(kg/ha)		
	75	150	225
	Without farmyard manure		
Sugar percentage	16·6	16·2	15·8
Juice purity	88·8	88·3	87·5
	With farmyard manure		
Sugar percentage	16·4	16·0	15·6
Juice purity	88·2	87·5	86·8

Collier[59] reporting 40 experiments with commercial fertilisers on farms drew attention to the small effect of a 'normal' dressing of nitrogen (0·96 cwt/acre) on sugar percentage. The decrease was much less than might be inferred from Adams'[8] experiments, which tested larger amounts of nitrogen. The normal dressing of nitrogen in Collier's experiments decreased juice purity by about 0·5%. Increasing the dressing to 1·44 cwt/acre resulted in a decrease of 0·3% in sugar and 0·4% in juice purity.

Boyd et al.[38] reviewed the effect of nitrogen fertiliser on sugar percentage (110 experiments) and juice purity (73 experiments). Small dressings of nitrogen had variable effects on sugar percentage but large dressings markedly decreased it (Table 94). All dressings of nitrogen decreased juice purity.

TABLE 94   *Table 13-7*

EFFECT OF NITROGEN FERTILISER ON SUGAR PERCENTAGE
AND JUICE PURITY
(after Boyd et al.[38])

	Years	Number of fields	N dressing (cwt/acre)			
			0[a]	0·6	1·2	1·8
				(kg/ha)		
			0[a]	75	150	225
Sugar percentage	1957–65	110	(17·3)	17·3	16·8	16·3
Juice purity	1960–65	73	(94·7)	94·5	93·9	93·2

[a] Not tested on all fields.

More detailed examinations of the effect of increasing the quantity of nitrate in sugar-beet plants were made by Last and Tinker.[217] They changed the leaf and petiole nitrate concentrations by using different quantities of fertiliser and determined the resultant effect on sugar percentage and juice purity in 1964 and 1965 (Fig. 27). Increases in nitrate nitrogen in petioles up to 700 ppm (sampled in June) had little effect on sugar percentage at harvest. On most fields there was a large decrease in sugar percentage when the nitrate nitrogen concentration was greater than 700–800 ppm. On average, a difference of 180 ppm in petioles corresponded to a difference of 1% in sugar in the roots at harvest. Sugar percentage and juice purity in the experiments were less in 1965 than 1964 but nitrate concentrations were greater in 1964. The authors stressed that sugar percentage and juice purity depend on many other factors and, although nitrate concentrations were related to sugar percentage and juice purity in single experiments, they found no dependable relationships between them for all experiments.

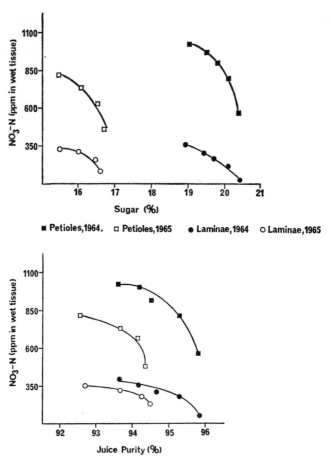

FIG. 27.   Sugar percentage, juice purity and the nitrate nitrogen concentration of petioles and laminae in 1964 and 1965.[217]

IN OTHER EUROPEAN COUNTRIES

Results from Ireland[234] confirm the findings in Great Britain. Results of 28 experiments showed that a small dressing of nitrogen (0·36 cwt/acre) had no effect on sugar percentage but further additions of fertiliser decreased sugar percentage by 0·1% for each 0·18 cwt/acre N. Even the smallest dressing of nitrogen decreased juice purity by 0·2%. Throughout the range tested (0–1·0 cwt/acre) juice purity decreased almost linearly (96·4–95·3%). Carolan,[50] also working in Ireland, found that nitrogen fertilisers caused large increases in the glutamine concentration of sugar beet and this partly accounted for the decrease in juice purity. Climatic conditions in Ireland are such that sugar beet is often still growing at harvest.

In France, Dubourg et al.[106] examined the effect of a very wide range of amounts of nitrogen (0–2·0 cwt/acre N) on sugar percentage and on the concentration of amino acids in the soluble fraction of sugar-beet roots. Sugar percentage was affected little by small dressings of nitrogen but large quantities decreased it linearly. All the nitrogen dressings increased the amino acid concentration.

Lüdecke and Nitzsche[231] investigated the effects of 'normal' and excessive amounts of fertiliser on sugar percentage and juice purity in Germany. Normal fertilising (1·0, 1·0 and 1·4 cwt/acre N, P_2O_5 and K_2O respectively) had little effect on quality but over-fertilising (3·2, 3·2 and 4·5 cwt/acre N, P_2O_5 and K_2O respectively) decreased it drastically, partly by retarding certain metabolic processes in autumn. Winner[378] also in Germany and Asselbergs et al.[13] in Holland described similar results.

In a pot experiment von Müller et al.[352] tested the influence of the $N:K_2O$ ratio on quality of sugar beet. When giving a large supply of nitrogen it was important to have a wide $N:K_2O$ ratio (1:3) for maximum quality. When the ratio was narrow (1:1) maturity was retarded and quality decreased due to an increase in glutamine concentration in the roots. Heistermann[166] confirmed these results, finding that a wide $N:K_2O$ ratio increased the dry matter content of the roots, and hence the sugar percentage.

IN THE USA
Ryser[297] made a study of the effect of nitrogen and other agronomic practices in sugar beet grown on farms in Oregon. Nitrogen decreased sugar percentage as shown below:

N dressing (cwt/acre)	%
below 0·99	16·29
1·00–1·50	16·08
1·51–2·00	16·05
2·01 and above	15·79

Delaying the harvest increased the sugar percentage and decreased the effect of nitrogen. In Minnesota[259] even a small nitrogen dressing caused a large decrease in sugar percentage and juice purity. Both decreased linearly throughout the range tested and the effect on sugar percentage was so great that, despite an increase in root yield, sugar yield was not increased. However, the sugar beet followed a legume, so the residue of nitrogen was presumably sufficient to satisfy the requirement of nitrogen by the sugar beet without further addition as fertiliser.

In California, Stockinger et al.[321] confirmed that different cropping systems influenced both yield and quality by their effect on the availability of soil nitrogen. Cropping systems which added organic matter or left residual N from large fertiliser applications increased yield and decreased sugar percentage and juice purity. With 3·75 cwt/acre N there was no significant difference between cropping systems. The nitrogen from organic sources had no advantage over inorganic fertiliser nitrogen.

Haddock et al.[144] in Utah investigated the effect of nitrogen fertiliser on sugar-beet quality in more detail. They measured a range of nitrogen-containing constituents in the sugar-beet root and by regression analysis showed that large concentrations of glutamine and ammonium nitrogen decreased quality most. Glutamine in particular was closely related to quality—where nitrogen fertiliser or other cultural factors increased glutamine concentration, quality always decreased. In a later report, Haddock et al.[145] established that nitrogen concentration in sugar-beet tissue is inversely related to sugar percentage and juice purity. Where sugar percentage and juice purity are small, top:root ratios are large.

Phosphorus

Adams[8] reported that phosphorus fertiliser had no effect on sugar percentage or juice purity of sugar beet in 41 experiments on commercial farms. Draycott and Cooke[80] found a small increase (+0·02%) in sugar percentage, but juice purity was not affected. In experiments on peaty soils, Tinker[333] reported that 0·33 cwt/acre P_2O_5 increased sugar percentage by 0·1 but had no effect on juice purity.

McDonnell et al.[234] described the effect of superphosphate (18% P_2O_5) on sugar percentage and juice purity of sugar beet in Ireland. Sugar percentage was increased considerably by the smallest dressing (Table 95) but larger amounts had no effect. Juice purity was increased significantly in the first year but had no effect in the second year. Gericke[131] found surprisingly large responses to phosphorus fertiliser in sugar percentage of sugar beet in Germany. It also improved the feeding value of the sugar-beet leaves, for protein concentration was increased considerably and oxalic acid concentration decreased by a third.

There are conflicting reports of the effect of phosphorus fertiliser

TABLE 95

EFFECT OF PHOSPHORUS FERTILISER ON SUGAR
PERCENTAGE AND JUICE PURITY IN IRELAND: MEAN OF 18
EXPERIMENTS
(after McDonnell et al.[234])

P dressing (cwt/acre P_2O_5)	(kg/ha P)	Sugar percentage	Juice purity (%)
0	0	16·53	95·80
0·73	40	16·68	95·88
1·46	80	16·68	95·91
2·19	120	16·70	95·97

on sugar-beet quality from the United States of America. In an experiment in Minnesota, Ogden et al.[259] reported a large and significant decrease in sugar percentage from large dressings, but juice purity was affected little, as seen below:

P_2O_5 (cwt/acre)	Sugar percentage	Juice purity %
0	16·55	81·79
1·8	16·32	81·87
3·6	16·11	81·85

However, in Kansas[174] phosphorus fertiliser had no effect on sugar percentage when tested at 0 and 1·10 cwt/acre P_2O_5.

It appears from the experiments in Great Britain and abroad that phosphorus fertiliser usually has a positive effect on sugar percentage. On soils with small concentrations of available phosphorus, it can be of the order of $+0·3\%$ or more. However, on soils naturally rich in phosphorus or where regular additions of fertiliser have been made, it is unlikely that the effect of even a large dressing will exceed $+0·1\%$. All the evidence suggests that phosphorus fertilisers have no effect on juice purity.

Potassium and sodium

Herron et al.,[174] in experiments in Kansas, tested various fertilisers and plant spacings on sugar-beet yield and quality. On average of other factors, potassium increased sugar percentage slightly but had no effect on juice purity. Winner[378] in Germany confirmed the increase in sugar percentage:

K_2O (cwt/acre)	Sugar percentage
0	18·1
0·8	18·4
1·6	18·4
2·4	18·7

Winner also tested nitrogen fertiliser but found there was no interaction with potassium in sugar percentage. Draycott and Cooke[80] reported a similar increase from potassium fertiliser in Great Britain. They also tested sodium fertiliser which increased sugar percentage by a similar amount to potassium. There was a large negative interaction between sodium and potassium in their effect on sugar percentage.

Contrary to these results, Simon et al[312] in Belgium found the potassium dressings decreased sugar percentage:

K_2O (cwt/acre)	Sugar percentage
1·6	17·09
3·2	16·72
4·8	16·84
6·4	16·54

Potassium and sodium are two serious impurities which decrease extraction of white sugar in the processing of the roots (see page 212). It is therefore expected that giving these elements in fertilisers increases their concentration in the root. However, Simon et al.[312] found that even an excess of potassium in Belgian soils did not adversely affect the quality of the root juice. In Great Britain, Draycott et al.[89] determined the sodium, potassium and α-amino nitrogen (another very serious impurity in sugar-beet juice) in crops which had been treated with a wide range of sodium (0–7·6 cwt/acre NaCl) and potassium (0–6·3 cwt/acre KCl) fertiliser. As expected, each fertiliser increased the concentration of that element in the roots by significant but, especially in the case of the large applications, surprisingly small amounts. What is also of interest, each decreased the concentration of α-amino nitrogen, which probably accounts for the small effect of potassium and sodium fertilisers on juice purity.

McDonnell et al.[234] reported experiments which confirm that potassium fertiliser increases sugar percentage but has no effect on juice purity:

K_2O (cwt/acre)	Sugar percentage	Juice purity %
0	16·35	95·87
1·1	16·67	95·95
2·2	16·77	95·89
3·2	16·81	95·84

Magnesium

Jorritsma[197] reported that magnesium fertiliser improved juice purity of sugar beet grown in Holland. Tinker[330] tested magnesium sulphate on sugar beet grown on fields where deficiency of the element was likely. On average of 17 fields, sugar percentage was increased by about 0·2% and juice purity also improved slightly. In 19 similar experiments[83] the average increase from 5 cwt/acre kieserite was 0·1 in sugar percentage, with no effect on juice purity. However, on fields where magnesium deficiency was severe (80% of the plants with symptoms on the leaves), sugar percentage was increased greatly (+0·5%) with an accompanying increase in juice purity (+0·2%).

Lachowski[213] grew a sugar-beet seed crop in soil with small exchangeable magnesium (34 ppm Mg) with and without 90 lb/acre magnesium sulphate. The magnesium improved the sugar percentage in the roots of the progeny, presumably due to an improvement in the magnesium concentration of the seed.

Effects of fertilisers on the keeping quality of sugar beet

In most sugar-beet growing countries, harvested roots are stored for varying periods before processing. Much of the crop in Great Britain is kept in storage heaps and clamps on farms, particularly that part of the crop harvested near the end of the growing season, and roots are also stored for relatively shorter periods at the factories (Martens and Oldfield[242] have reviewed the storage of sugar beet in 16 European countries). Thus it is important to know the effects of cultural practices on the storage properties of the roots; however, surprisingly few investigations have been made.

An early study of the effect of phosphate on the keeping quality of sugar beet was made in Utah by Larmer.[215] He grew the crop under conditions of adequate and inadequate phosphorus supply on fields known to be deficient in the element. Roots from plots receiving superphosphate or a complete fertiliser showed less decay than roots from unfertilised plots. In these tests the roots were kept in coarse-meshed sacks and exposed to conditions of commercial storage heaps. Besides the reduction of loss by decay, the phosphate fertiliser decreased the loss of sugar due to transpiration.

In another experiment Larmer found indications that nitrogen fertiliser and farmyard manure improved keeping quality. Also in Utah, Stout and Smith[324] grew sugar beet with and without a commercial dressing of fertiliser (exact details of the fertiliser are not given). It had little effect on the respiration rate of the stored roots but there was no response in yield to the fertiliser. Large roots did, however, respire more slowly than an equal weight of smaller roots.

In Colorado, Gaskill[130] tested 0, 0·45 and 1·35 cwt/acre N, 0 and 0·45 cwt/acre K_2O and 0, 0·90 and 1·80 cwt/acre P_2O_5. Roots from the field experiment were stored at 45°F and 65°F for three and four months, and those grown with nitrogen kept better at 45°F than those without. At 65°F there were no significant differences attributable to nitrogen. Effects of potassium and phosphorus were small, which the author considered not surprising, as there was a negligible yield response to potassium and relatively little to phosphorus.

It seems from this sparse experimental evidence that a commercial dressing of nitrogen fertiliser is likely to improve the keeping quality of roots. Boron applications are also important where the element is short, for roots of deficient plants deteriorate rapidly in storage. Where soils have little available phosphorus, fertiliser is also likely to improve keeping quality but where yield responses are small, quality will change little. The effects of other major and minor nutrient elements remain to be determined, as does the effect of excessive dressings of fertiliser.

Conclusions

NITROGEN

The mean effect of nitrogen fertiliser on sugar percentage and juice purity is as shown in the following table:

N (cwt/acre)	Sugar percentage	Juice purity %
0	17·0	93·1
0·60	16·8	92·6
1·20	16·5	92·2
1·80	16·0	91·6

The first 0·60 cwt/acre N decreases sugar percentage by only 0·2% but more has a progressively larger effect, and 1·80 cwt/acre decreases it by 1·0%. Nitrogen decreases juice purity linearly, each 0·60 cwt/acre N by about 0·5%.

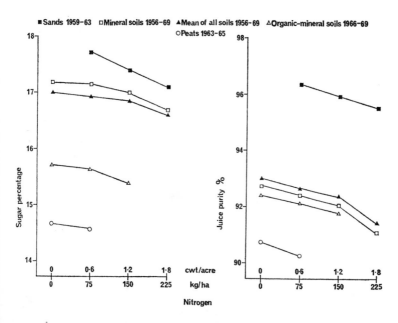

FIG. 28. Effect of nitrogen fertiliser and soil type on sugar percentage and juice purity.[103]

Nitrogen has a surprisingly similar effect on widely differing soils (Fig. 28). Draycott et al.[103] showed that roots of sugar beet grown on peat soils had an average sugar percentage of only 14·7 and juice purity of 90·8%, whereas averages on sandy soils were 17·0 and 96·0% respectively.

PHOSPHORUS

Phosphorus fertiliser increases sugar percentage slightly on severely deficient soils but on moderately fertile fields neither increases nor decreases it. Juice purity is not affected by phosphorus fertiliser on any soil.

POTASSIUM, SODIUM AND MAGNESIUM

On average, each of these cations improves sugar percentage by 0·1 to 0·2%. There are conflicting reports of the effects of potassium and sodium fertilisers on juice purity. Some[312,174,103] report that potassium has no effect on juice purity and some[103] that sodium, too, has no effect. However, Carruthers et al.[51] found that sodium decreased juice purity. As both elements are serious impurities in the root juice, increasing their concentration in the plant would be expected to decrease purity. However, increasing the concentration of sodium or potassium in the root probably decreases the concentration of some other impurity.

EFFECT OF CHANGES IN QUALITY ON THE VALUE OF THE CROP

In most countries, growers are paid either in proportion to the total amount of sugar in the roots (the product of root weight and sugar percentage) or on a scale paying more (or less) than pro-rata for increases (or decreases) in sugar percentage. The latter method is used in Great Britain; the payment at present is £7·60/ton at 16%

TABLE 96

CHANGES IN THE VALUE OF THE CROP FROM INCREASES IN NITROGEN FERTILISER APPLICATION IN GREAT BRITAIN
(after Draycott et al.[103])

N dressing	Root yield	Sugar	Payment		Total sugar	Juice purity	Extractable white sugar
(cwt/acre)	(ton/acre)	(%)	(£/ton roots)	(£/acre)	(cwt/ acre)	(%)	(cwt/acre)
0·0	12·34	17·30	8·45	104·27	42·7	93·1	33·5
0 to 0·6	+3·40	−0·10	−0·05	+27·95	+11·4	−0·5	+8·8
0·6 to 1·2	+1·24	−0·45	−0·20	+7·02	+2·7	−0·4	+0·9
1·2 to 1·8	+0·06	−0·63	−0·35	−5·48	−1·9	+0·6	−2·8
(kg/ha)	(t/ha)	(%)	(£/t roots)	(£/ha)	(t/ha)	(%)	(t/ha)
0	30·99	17·30	8·59	257·44	5·36	93·1	4·20
0 to 75	+8·54	−0·10	−0·05	+69·01	+1·43	−0·5	+1·10
75 to 150	+3·11	−0·45	−0·20	+17·33	+0·34	−0·4	+0·11
150 to 225	+0·15	−0·63	−0·36	−13·53	−0·24	−0·6	−0·35

sugar, $\pm£0\cdot50$ per 1% above or below 16%, so that slightly more is paid for 17 tons of roots at 18% than 18 tons at 17% sugar.

Table 96 shows how nitrogen fertiliser affects the value of the crop in Great Britain per ton of roots and per acre. Payment per ton of roots is greatly decreased by nitrogen, simply because sugar percentage is decreased, but payment per acre shows that optimum dressing is about $1\cdot00$ cwt/acre. In countries where payment is for total sugar the optimum dressing is slightly more. When the amount of extractable white sugar is estimated, the juice purity needs to be taken into account, and Table 96 shows that for maximum extractable white sugar/acre, less nitrogen ($0\cdot20$–$0\cdot30$ cwt/acre) fertiliser should be used.

Chapter 14

Nutrient Requirement of the Seed Crop

Sugar beet is a biennial plant, only producing seed after vernalisation and when grown with suitable day length. The seed crop is produced either by sowing the mother seed in summer of the first year and transplanting the 'stecklings' the following spring, or, increasingly, by direct (*in situ*) sowing without transplanting. The former system is the most common method of multiplication for European countries except Denmark and England. Direct sowing is the only method of production in North America.[307] The direct-sown crop is sometimes grown under a cover crop which protects the sugar-beet plants from pests and diseases.

The various methods of seed production allow many alternative manurial practices for the crop. However, as only 3 000 acres of seed crop are grown in England each year, relatively little attention (compared with the root crop) has been given to investigating its nutrient requirement. The experimental results available are summarised below, supplemented by experience of seed crop nutrient needs reported from North America and elsewhere.

The seed crop in Europe

NITROGEN

Transplanted crops
Ellerton[115] in one of the first experiments on the manurial requirement of the transplanted seed crop in England, tested 0, 2 and 4 cwt/acre ammonium sulphate given in March with all combinations of 0 and 2 cwt/acre in June when the plants began to flower. Giving the nitrogen in March increased yield considerably—2 cwt/acre N by 16% and 4 cwt/acre by 25%. Nitrogen in June was less effective—2 cwt/acre increased yield by only 8%. Nitrogen given in March did not affect the cluster size but increased the germination percentage.

In the same experiments the effect of 'topping'—that is, the removal of the growing point to encourage lateral shoots to develop—was also tested. Although on average nitrogen did not affect date of flowering, there was an interaction between nitrogen applied in March and topping, for nitrogen appeared to accelerate the flowering of plants topped early. The author considered that this may in part have accounted for the improvement in germination caused by nitrogen application.

Ellerton[116] in a later experiment grew transplanted crops of three varieties with 8 cwt/acre 12:12:18 ($\%N:P_2O_5:K_2O$) compound plus 0·60 cwt/acre N as 'Nitro-Chalk' worked into the soil and tested an additional 0·60 cwt/acre N as sodium nitrate top dressing. The basal dressing was sufficient for maximum yield of all the varieties. Mann and Barnes[241] experimented with the spring-transplanted seed crop at Woburn. They found that an application of 0·4 cwt/acre N increased yields of seed from 11·5 to 15·2 cwt/acre in 1942 and from 9·8 to 11·5 cwt/acre in 1943. Germination was depressed by the nitrogen application.

Ellerton[117] grew a Bush 'E' seed crop with 4 cwt/acre of $5:12\frac{1}{2}:12\frac{1}{2}$ basal fertiliser and tested the value of an additional 4 cwt/acre ammonium sulphate (21% N). The experiment was sown in 1957 under a pea cover crop and given the basal fertiliser, the nitrogen treatment being applied in the following spring. The yield of seed without the nitrogen dressing was 20·0 cwt/acre and giving the nitrogen increased yield to 23·1 cwt/acre. The cost of the fertiliser was about £2 and the extra seed was worth about £17. The profitability of the application was large enough to suggest that even more nitrogen should be given. Nitrogen fertiliser did not, however, affect the average weight of each cluster nor the percentage of seeds germinating in laboratory tests.

Sneddon[317] investigated response to nitrogen by crops transplanted in spring at Cambridge. Nitrogen top-dressing (1·00 or 2·00 cwt/acre N) increased seed yield in only one year out of three, presumably because the base dressing (not stated) and the soil nitrogen were sufficient. Also, both years when nitrogen did not increase yield were wet, which prolonged vegetative growth into September, causing any increase in yield from the extra nitrogen to be delayed beyond a practical harvest date. Where nitrogen was given, the large clusters at the base of the raceme were slow to ripen and the small, late-formed clusters at the apex were insufficiently developed to add to the yield of seed. In the dry year, even with nitrogen the racemes had ceased extensive growth and turned brown early and the crop was harvested in August. Nitrogen increased seed yield by 42% but straw yield by only 27%.

Direct-sown crops

Sneddon[317] compared two times of application of top-dressing of 1·00 cwt/acre N on direct-sown crops. Nitrogen was given either before bolting commenced or after the plants had produced maximum extension and had developed flower buds. The nitrogen greatly increased yield in all three years of the experiment. Applying the nitrogen before bolting was slightly more effective than when given before flowering. Nitrogen only affected the seed size in one experiment, when the proportion of clusters over $\frac{10}{64}$ in was 40% compared with 35% for plots given no nitrogen top-dressing. Nitrogen, however, consistently decreased germination from about 75% to 67%.

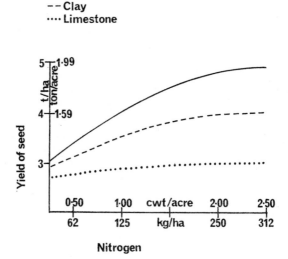

FIG. 29.   Nitrogen fertiliser and the yield of seed on three contrasting soils.[226]

Scott[308] described eight experiments in four areas testing nitrogen fertiliser applied in late February or in March/April of the harvest year on direct-sown crops. Two experiments were on deep organic silty clay loam soil in Lincolnshire, two on a clay loam in Northamptonshire, three on shallow Cotswold Limestone brash soil and one on Bunter Sandstone soil in Nottinghamshire. Although seed yields were similar without nitrogen on all fields, the response to nitrogen was much greater on the deep silty clay than on the shallow limestone soil. Not only did nitrogen nearly double yield on the silt but more nitrogen fertiliser was needed for maximum yield. The crop on the clay soil gave an intermediate response (Fig. 29). Scott concluded that the seed crop on deep organic silt responded to more nitrogen than on the shallow limestone soils because water limited yield on the latter. Nitrogen fertiliser never decreased yield significantly. In the experiment on Bunter Sandstone, the yields of seed from 0·7 to 1·4 cwt/acre N were similar; splitting the dressing so that two-thirds was given in April and the rest in June had no advantage over giving all in April.

Increasing the fertiliser supply increased yield because plants produced more clusters, not larger or heavier ones. Only in one experiment, where the crop was affected by adverse weather conditions, did nitrogen decrease germination percentage of the seed produced from 82 to 75%. Where the weather was good and the plots harvested at the normal time (August/September), nitrogen did not depress germination.

Scott[306] described an experiment to test the effect of 0, 0·40, 0·80, 1·20 and 1·60 cwt/acre N for an *in situ* crop. The nitrogen increased seed yield (Table 97) up to the largest amount tested. However,

TABLE 97

THE EFFECT OF NITROGEN FERTILISER GIVEN IN SPRING
TO AN *in situ* SEED CROP
(after Scott[306])

N dressing (cwt/acre)	0	0·40	0·80	1·20	1·60
(kg/ha)	0	50	100	150	200
Seed yield (cwt/acre)	21·9	22·4	23·0	23·0	25·4
(t/ha)	2·75	2·81	2·89	2·89	3·19
Germination percentage	81·5	82·3	81·3	81·3	74·5

ripening was delayed and germination percentage was decreased by the largest dressing.

PHOSPHORUS

In the experiment described above by Ellerton,[117] in addition to testing nitrogen, 0 and 4 cwt/acre of superphosphate (18% P_2O_5) was tested. It did not affect yield, cluster weight or germination percentage. Scott[308] tested 0·9 cwt/acre of both P_2O_5 and K_2O given together to direct-sown crops in four locations. All the soils had 'low' phosphorus status and 'medium' or 'high' potassium status. The fertiliser only increased yield in one experiment, by about

4 cwt/acre. The response to nitrogen was not affected by supplying phosphorus and potassium.

POTASSIUM

Ellerton[117] tested 0 and 2 cwt/acre potassium chloride (60% K_2O) on an *in situ* seed crop grown under peas. The potassium was given in spring of the second year but did not affect yield or quality of the seed.

SODIUM

As with the root crop, sugar-beet seed yields are increased by this element, which acts in a somewhat similar way to (and partially replaces the need for) potassium. Mann and Barnes[241] reported increased yield from sodium in early experiments and Longden[226] has recently confirmed the benefit of giving 3 cwt/acre NaCl in the seedbed of open-sown *in situ* crops.

MAGNESIUM

No experiments have been made with magnesium for the seed crop in England, but recent experiments with the root crop show that it is important to give the element on soils with less than 50 ppm; large responses were on soils with less than 25 ppm.[90] Lachowski[213] in Poland grew sugar beet in the field and in pots of a podsolic soil considered to be deficient in magnesium (34 ppm Mg). Magnesium sulphate decreased the amount of water used by the plants and the characteristic was maintained in the following generation. Giving magnesium to the seed crop increased the uptake of nitrogen, phosphorus, potassium and magnesium in the seed but decreased the amount of sodium in the seed. Magnesium gave an important increase in yield of 9–16% and improved germination by 5–11%. It also increased the sugar percentage and phosphorus, potassium and sodium concentration in leaves of plants produced from the seed.

BORON

Brenchley and Watson[41] transplanted sugar-beet plants showing slight or severe heart-rot symptoms to containers of acid-washed sand and gave various amounts of boric acid. In the absence of boron the characteristic symptoms appeared; the apices of the stems and the flower buds blackened and died. Plants with few or no symptoms present before transplanting also developed the necrosis in the seeding year. Giving boric acid completely removed symptoms of heart rot. Where heart rot was originally severe the main axis was killed but healthy lateral shoots were produced.

Effects of fertiliser on seed composition

NUTRIENT CONCENTRATION

Longden[226] reported on the effect of fertiliser on the chemical composition of the seeds produced. Nitrogen at 1·5, 2·0 and 2·5 cwt/acre gave seed with 2·04, 2·13 and 2·26% N respectively in the dry matter. Phosphorus fertiliser did not affect the concentration of the element in the dry matter and when samples of the seeds were sown in the field there were no effects on emergence or seedling weight. Longden concluded that it was unlikely that major mineral element deficiencies occur in sugar-beet seed in this country.

AMOUNT OF NUTRIENTS IN THE SEED CROP

Longden[226] reported that seed contains 2·0% N, 0·45% P and 2·0% K in the dry matter. Leaves contain 2·7% N, 0·39% P and 2·9% K and roots 1·7% N, 0·36% P and 1·7% K. For a 20 cwt/acre crop of seed with 40 cwt/acre straw and 8 cwt/acre roots the crop at harvest contains 1·6 cwt/acre N, 0·62 cwt/acre P_2O_5 and 2·06 cwt/acre K_2O.

Seed crop in North America

Pultz[282] made experiments with direct-sown seed crops in Utah and New Mexico to determine response to nitrogen. 6·7 cwt/acre ammonium sulphate was given either in September of the sowing year or March of the harvest year or in two dressings of 3·4 cwt/acre on both occasions. The nitrogen increased yield greatly, as below:

	Yield of seed (cwt/acre)	Germination (%)
Autumn	32·1	74
Autumn and spring	32·1	76
Spring	34·0	80
None	24·1	76

There were no consistent differences in quality of the crop.

Chemical analysis showed that the concentrations of carbohydrates and nitrogenous compounds in the roots were maximal during winter dormancy. After seed stalk development began in spring, sugars and nitrogen were rapidly withdrawn from the roots. The loss of nitrogen from the roots was only slightly affected by fertiliser treatment and continued until most of the nitrogen had been withdrawn.

However, the amount of sugar removed from the root depended on the supply of nitrogen from the soil. When nitrogen was continuously available during the fruiting period, the sugar percentage in the root decreased steadily until the seed matured and this resulted in a long period of flower production and a large seed yield. Where nitrogen became a limiting factor during the fruiting period the loss of sugar from the roots ceased, flowering stopped prematurely, and seed yield from such plants was small.

Campbell[48] reviewed sugar-beet seed production in Oregon on fertile, deep loam and silt soils. The crop was direct-sown and irrigated in the first year. Heavy winter rainfall leached nitrogen, and large dressings were applied in the spring either as soon as ground conditions permitted or from aircraft. At sowing, the fertiliser given contained 0·70 cwt/acre N, 0·90 cwt/acre P_2O_5, 0·18 cwt/acre K_2O, 0·45 cwt/acre S and 5 lb/acre each of boron and magnesium. In spring 0·70 cwt/acre N was given, plus another 0·70 cwt/acre in summer.

Snyder[318] compared the vigour and percentage germination of seeds from crops grown with various amounts of nitrogen. The amount of nitrogen supplied to the parent plants had no pronounced effect and it was concluded that the best nitrogen dressing was the one which gave the largest yield of seed. Pendleton[270] also tested nitrogen, and phosphorus and potassium on seed yield and germination. Seed yield without fertiliser was small and germination was poor. 1·30 cwt/acre N more than doubled yield and 2·70 cwt/acre N produced a further increase. Nitrogen and phosphorus increased germination percentage; together they increased it by 8%.

Tolman[335] experimented with nitrogen and phosphorus fertilisers for the seed crop in Utah. Applications of 5·40 cwt/acre of ammonium sulphate in spring increased yield of seed by about 6 cwt/acre and 2·7 to 3·6 cwt/acre triple superphosphate in autumn increased it by 9 cwt/acre. Overpeck and Elcock[264] working in New Mexico found that spring applications of 0·45 cwt/acre ammonium sulphate gave striking increases in yield and 9 ton/acre farmyard manure given before sowing was also very beneficial.

Pendleton et al.[267] described the soils best suited for sugar-beet seed production in Oregon. The crop needed a deep fertile soil with a large water-holding capacity, without restriction to root development. 0·45 to 0·70 cwt/acre N given in autumn of the seeding year was needed for maximum yield, with at least half as ammonium sulphate to ensure that the plants were not short of sulphur. Consistent responses to phosphorus and potassium were not general, but some areas needed boron as 25–30 lb/acre borax and others responded to 0·80 cwt/acre elemental sulphur.

Pendleton[269] in later experiments compared different forms of nitrogen for the sugar-beet seed crop grown on a sandy loam. Ammonium sulphate, nitrate and phosphate, sodium and calcium nitrate and urea were compared. They were all side-dressed in three equal increments of 0·90 cwt/acre N, in autumn, in April and in May. Basal dressings of 25 lb/acre borax and 1·10 cwt/acre gypsum were applied to correct boron and sulphur deficiencies. No phosphorus or potassium were given for the soils were well-supplied. The smallest yield was from plots given ammonium sulphate alone, the nitrates giving slightly more yield. Germination was not affected by any of the forms of nitrogen.

Fertiliser-use on the seed crop

Scott[307] made a survey of sugar-beet growing in Europe and North America during the last ten years. Table 98 shows the amount of

TABLE 98

FERTILISER USED FOR SEED CROPS GROWN BY THE INDIRECT METHOD: MEANS FROM 13 EUROPEAN COUNTRIES
(after Scott[307])

	1st Year	2nd Year
	(cwt/acre)	
N	0·61	1·04
P₂O₅	0·77	1·12
K₂O	1·06	1·71
	(kg/ha)	
N	76	131
P	42	61
K	110	177

fertiliser used for crops grown by the indirect method. In most countries the seed crop in the harvest year received a similar dressing of fertiliser to the root crop. Although the nitrogen dressing was about 0·80–1·28 cwt/acre N, the phosphorus and potassium dressings varied widely to suit local soil conditions. Amounts of fertiliser given for direct-sown crops in North America are shown in Table 99. In USA only nitrogen was applied in the second year, and most was given in the Southern states.

TABLE 99
FERTILISER APPLICATION
FOR DIRECT-SOWN SEED
CROPS IN NORTH AMERICA
(after Scott[307])

	1st Year	2nd Year
	(cwt/acre)	
N	0·58	2·06
P_2O_5	0·79	2·40
K_2O	0·36	1·21
	(kg/ha)	
N	73	258
P	43	131
K	37	125

Conclusions

TRANSPLANTED CROPS

Little is known about the amount of fertiliser needed to produce good stecklings for transplanting. Presumably their needs are similar to the sugar-beet root crop, thus on moderately fertile soils 1·00 cwt/acre N, 0·50 cwt/acre P_2O_5 and 1·00 cwt/acre K_2O given in the seedbed would probably be sufficient. Numerous experiments

TABLE 100
AVERAGE OPTIMAL FERTILISER DRESSINGS FOR DIRECT-SOWN
SEED CROPS IN ENGLAND
(after Longden[226])

	Undersown crops		Open-sown crops	
	First year (as top dressing in autumn)	Second year (as top dressing in spring)	First year (in seedbed)	Second year (as top dressing in spring)
	(cwt/acre)			
N	0·70	1·40	0·60	1·30
P_2O_5	0·90	0·20	0·90	0·70
K_2O	1·40	0·30	1·10	0·70
NaCl	—	—	3·00	—
	(kg/ha)			
N	88	176	75	163
P	49	11	49	38
K	146	31	115	73
Na	—	—	150	—

have been made to investigate the needs of the crop in the second year. Longden[226] considered 1·20 cwt/acre N, 0·90 cwt/acre P_2O_5 and 1·40 cwt/acre K_2O plus 3·00 cwt/acre NaCl was optimal in Great Britain.

DIRECT-SOWN CROPS

Table 100 summarises optima for direct-sown crops. Manures applied when the crop is undersown are usually those required by the cover crop, commonly barley. Manuring of the sugar-beet seed crop usually starts after the barley is harvested, with a top-dressing of compound fertiliser supplying on average 0·70, 0·90 and 1·40 cwt/acre N, P_2O_5 and K_2O respectively, plus 3·00 cwt/acre NaCl. About the same amount should be given in the seedbed of open-sown crops. In the second year, 1·20 to 1·30 cwt/acre N is needed as a top-dressing, which often increases yield greatly. Although phosphorus and potassium are often given, there is no experimental evidence to support their use for either undersown or open-sown crops.

Appendix 1

Conversion Factors

To Convert	To	Multiply by	Reciprocal
Inches (in)	Centimetres (cm)	2·54	0·394
Yards (yd)	Metres (m)	0·914	1·09
Square yards (yd^2)	Square metres (m^2)	0·836	1·20
Acres (4 840 yd^2)	Hectares (ha)	0·405	2·47
Pounds (lb)	Kilogrammes (kg)	0·454	2·20
Hundredweights (cwt) (112 lb)	Kilogrammes (kg)	50·8	0·019 7
Tons (2 240 lb)	Tonnes (t)	1·02	0·984
Short tons (2 000 lb)	Tonnes (t)	0·911	1·10
lb/acre	kg/ha	1·12	0·892
cwt/acre	kg/ha	126	0·007 97
ton/acre	t/ha	2·51	0·398
P_2O_5	P	0·436	2·29
K_2O	K	0·830	1·20
CaO	Ca	0·715	1·40
MgO	Mg	0·603	1·66
Na_2O	Na	0·742	1·35
NaCl	Na	0·393	2·70
cwt/acre:	lb/acre:		
P_2O_5	P	49·0	0·020 4
K_2O	K	92·6	0·010 8
cwt/acre:	kg/ha		
P_2O_5	P	54·8	0·018 3
K_2O	K	104	0·009 64
NaCl	Na	50·0	0·020

Terminology

Field capacity: Amount of water in soil after rain has stopped and after drainage has ceased.

Juice purity: The concentration of sugar (sucrose) percent of total soluble solids in fresh roots determined by the Carruthers and Oldfield method (page 201).

ppm: Parts per million.

Root yield: The weight of clean fresh roots (topped in Great Britain at the level of the lowest leaf scar) per unit area.

Sugar percentage: The concentration of sugar (sucrose) percent of fresh roots determined by the Sachs–Le Docte method (page 201).

Sugar yield: The weight of sugar in the crop per unit area, *i.e.* the product:

$$\text{Fresh root yield} \times \frac{\text{Sugar percentage}}{100}$$

Tops yield: The weight of clean fresh laminae, petioles and crowns (cut in Great Britain from the root at the level of the lowest leaf scar) per unit area.

Wilting point: Amount of water in soil at which plants just wilt and do not recover in a saturated atmosphere.

References

1. ADAMS, S. N. (1959). Manuring next year's sugar beet, *Br. Sug. Beet Rev.* **28**, 77–8.
2. ADAMS, S. N. (1960). The value of calcium nitrate and urea for sugar beet and the effect of late nitrogenous top dressings, *J. agric. Sci., Camb.* **54**, 395–8.
3. ADAMS, S. N. (1961). The effect of time of application of phosphate and potash on sugar beet, *J. agric. Sci., Camb.* **56**, 127–30.
4. ADAMS, S. N. (1961). The effect of sodium and potassium on sugar beet on the Lincolnshire limestone soils, *J. agric. Sci., Camb.* **56**, 283–6.
5. ADAMS, S. N. (1961). The effect of sodium and potassium fertiliser on the mineral composition of sugar beet, *J. agric. Sci., Camb.* **56**, 383–8.
6. ADAMS, S. N. (1961). The rôle of sodium in manuring sugar beet in Great Britain, *Proc. Int. Inst. Sug. Beet Res. XXIV Congr., Brussels,* 311–16.
7. ADAMS, S. N. (1961). The manuring of sugar beet, *Chemy Ind.* 564–6.
8. ADAMS, S. N. (1962). The response of sugar beet to fertiliser and the effect of farmyard manure, *J. agric. Sci., Camb.* **58**, 219–26.
9. AGERBERG, L. S. (1969). [Results of three different cropping systems at different plant nutrient levels], *Lantbr Högsk. Meddn Ser. A.* **117**, 1–42.
10. ALBASAL, N., DOR, Z., CARMELI, R. and KAFKAFI, U. (1970). Growth of sugar beet (*var. polyrave*) in relation to petiole nitrate content, *Exp. Agric.* **6**, 151–5.
11. ALBBECHT, W. A. (1970). Nutritional rôle of calcium in plants. Part I. Prominent in the non-legume crops, sugar beet, *Pl. Soil* **33**, 361–82.
12. ALLEN, M. (1968). A progress report on experiments with magnesium applied in various ways to some fruit soils. NAAS 'Open' Conf. of Soil Scientists—'Residual value of applied nutrients'.
13. ASSELBERGS, C. J., VAN DER POEL, P. W., VERHAART, M. L. A. and DE VISSER, N. H. M. (1960). Rohsaftgewinnung im laboratorium zum Studium des Techinschen wertes der Zuckerrübe, *Proc. XIth Sess. Comm. Int. Tech. Suc.* 78–91.
14. ATKINSON, H. J., GILES, G. R. and DESJARDINS, J. G. (1954). Trace element content of farmyard manure, *Can. J. agric. Sci.* **34**, 76–80.
15. BAIRD, B. L., BONNEMANN, J. J. and RICHARDS, A. W. (1954). The use of chemical additives to control soil crusting and increase emergence of sugar beet seedlings, *Proc. Am. Soc. Sug. Beet Technol.* **8**, 136–42.
16. BALDWIN, C. S. and STEVENSON, C. K. (1969). The effect of nitrogen on yield, percent sucrose, and clear juice purity of sugar beets, *J. Am. Soc. Sug. Beet Technol.* **15**, 522–7.
17. BASSEREAU, D. (1970). Sugar plants: beet and cane, *Agron. trop., Paris* **25**, 928–34.

18. BAVER, L. D. and FARNSWORTH, R. B. (1940). Soil structure effects in the growth of sugar beets, *Soil Sci. Soc. Am. Proc.* **5**, 45–8.
19. BERGER, K. C. (1950). Sugar beet fertilisation in Wisconsin, *Proc. Am. Soc. Sug. Beet Technol.* **6**, 440–4.
20. BERNSHTEIN, B. I. and OKANENKO, A. S. (1966). [Effect of potassium deficiency on photosynthesis, respiration and phosphorus metabolism in ontogeny of sugar beet], *Fiziologiya Rast.* **13**, 629–39.
21. BIRCH, J. A., DEVINE, J. R. and HOLMES, M. R. J. (1966). Field experiments on the magnesium requirement of cereals, potatoes and sugar beet in relation to nitrogen and potassium application, *J. Sci. Fd. Agric.* **17**, 76–81.
22. BJÖRLING, K. (1954). Yellowing in beets caused by magnesium deficiency, *Socker* **8**, 147–56.
23. BLAKE, G. R., OGDEN, D. B., ADAMS, E. P. and BOELTER, D. H. (1960). Effect of soil compaction on development and yield of sugar beets, *J. Am. Soc. Sug. Beet Technol.* **11**, 236–42.
24. BLAND, B. F. (1957). The use of nitrogen for sugar beet on a heavy loam soil in Norfolk, *Expl. Husb.* **2**, 33–6.
25. BLENCOWE, J. W. and TINSLEY, T. W. (1951). The influence of density of plant population on the incidence of yellows in sugar beet crops, *Ann. appl. Biol.* **38**, 395–401.
26. BLOOD, J. W. (1957). Chemical aspects of soil advisory work, *J. Sci. Fd. Agric.* **8**, 645–53.
27. BOAWN, L. C. and VIETS, F. G. JR. (1956). Zinc fertiliser tests on sugar beets in Washington, *J. Am. Soc. Sug. Beet Technol.* **9**, 212–16.
28. BOAWN, L. C., VIETS, F. G. JR., NELSON, C. E. and CRAWFORD, C. L. (1961). Yield and zinc content of sugar beets as affected by nitrogen source, rate of nitrogen and zinc application, *J. Am. Soc Sug. Beet Technol.* **11**, 279–86.
29. BOITEAU, R. (1972). Private communication.
30. BOLTON, E. F. and AYLESWORTH, J. W. (1968). Effect of soil physical condition and fertility on yield of sugar beets on a Brookston clay soil, *J. Am. Soc. Sug. Beet Technol.* **14**, 664–70.
31. BOLTON, J. and PENNY, A. (1968). The effects of potassium and magnesium fertilisers on yield and composition of successive crops of ryegrass, clover, sugar beet, potatoes, kale and barley on sandy soil at Woburn, *J. agric. Sci., Camb.* **70**, 303–11.
32. BOYD, D. A., GARNER, H. V. and HAINES, W. B. (1957). The fertiliser requirements of sugar beet, *J. agric. Sci., Camb.* **48**, 464–76.
33. BOYD, D. A. (1959). The effect of farmyard manure on fertiliser responses, *J. agric. Sci., Camb.* **52**, 384–91.
34. BOYD, D. A. (1961). Current fertiliser practice in relation to manurial requirements, *Proc. Fertil. Soc.* No. 65.
35. BOYD, D. A., CHURCH, B. M. and HILLS, M. G. (1961). Fertiliser practice in England and Wales. I. General features of fertiliser consumption 1956–7, *Emp. J. exp. Agric.* **29**, 35–44.
36. BOYD, D. A. (1968). Experiments with ley and arable farming systems, *Rep. Rothampstead Exp. Stn. for* 1967, 316–31.
37. BOYD, D. A. (1970). Private communication.
38. BOYD, D. A., TINKER, P. B. H., DRAYCOTT, A. P. and LAST, P. J. (1970). Nitrogen requirement of sugar beet grown on mineral soils, *J. agric. Sci., Camb.* **74**, 37–46.
39. BRANDENBURG, E. (1931). Die Herz- und Trockenfäule der Rüben als Bormangelerscheinung, *Phytopath. Z.* **3**, 499–517.
40. BRANDENBURG, E. (1939). Ueber die Grundlagen der Boranwendung in der Landwirtschaft, *Phyopath. Z.* **12**, 1–112.

41. BRENCHLEY, W. E. and WATSON, D. J. (1937). The influence of boron on the second year's growth of sugar beet affected with heart rot, *Ann. appl. Biol.* **15**, 494–503.

42. BROWN, A. L., HILLS, F. J. and KRANTZ, B. A. (1968). Lime, P, K and Mn interactions in sugar beets and sweet corn, *Agron. J.* **60**, 427–9.

43. BROWN, R. J. (1943). Sampling sugar beet petioles for measurement of soil fertility, *Soil Sci.* **56**, 213–222.

44. BROWN, R. J. and WOOD, R. R. (1952). Improvement of processing quality of sugar beets by breeding methods, *Proc. Am. Soc. Sug. Beet Technol.* **7**, 314–18.

45. BRUMMER, V. (1966). [Effect of autumn and spring applications of fertiliser on sugar beets], *Maatalous Koetoim.* **20**, 91–100.

46. BUNTING, A. H. (1963). Experiments on organic manures, 1942–9, *J. agric. Sci., Camb.* **60**, 121–40.

47. CAMPBELL, R. E. and VIETS, F. G. JR. (1967). Yield and sugar production by sugar beets as affected by leaf area variations induced by stand density and nitrogen fertilisation, *Agron. J.* **59**, 349–54.

48. CAMPBELL, S. C. (1968). Sugar beet seed production in Oregon, USA, *J. int. Inst. Sugar Beet Res.* **3**, 165–74.

49. CAPITAINE, R. C. (1965). Recommendations sur l'emploi des engrais en Tunisie, *Bull. Inst. natn. Rech. agron. Tunisie* **7**, 4.

50. CAROLAN, R. J. (1960). Nonsugars in factory juices with special reference to effective alkalinity, *Proc. XIth Sess. Comm. Int. Tech. Suc.* 203–20.

51. CARRUTHERS, A., OLDFIELD, J. F. T. and TEAGUE, H. J. (1956). A comparison of the effects on juice quality of nitrate of soda, sulphate of ammonia and salt, *Proc. Int. Inst. Sug. Beet Res. XIX Congr., Brussels.*

52. CARRUTHERS, A. and OLDFIELD, J. F. T. (1961). Methods for the assessment of beet quality, *Int. Sug. J.* **63**, 72–4, 103–5, 137–9.

53. CARRUTHERS, A., OLDFIELD, J. F. T. and TEAGUE, H. J. (1962). Assessment of beet quality. Paper presented to the XVth Ann. Tech. Conf. Br. Sug. Corp., pp. 1–28.

54. CHARLESWORTH, R. R. (1967). The effect of applied magnesium on the uptake of magnesium by, and on the yield of, arable crops, *Tech. Bull. Minist. Agric. Fish. Fd* **14**, 110–25.

55. CHRISTMANN, J. (1963). Results of trials on the application of nitrogenous fertilisers to sugar beet conducted by the Institut Technique de la Betterave from 1954–1962, *Proc. Int. Inst. Sug. Beet Res. XXVI, Congr., Brussels.*

56. CHURCH, B. M. (1952). Recent trends in fertiliser practice in England and Wales. Part II. The use of fertilisers in cereals, root crops, and grassland, *Emp. J. exp. Agric.* **80**, 257–70.

57. CHURCH, B. M. and WEBBER, J. (1971). Fertiliser practice in England and Wales: a new series of surveys, *J. Sci. Fd. Agric.* **22**, 1–7.

58. COHEN, A. (1972). Private communication.

59. COLLIER, P. A. (1967). I.C.I. Fertiliser trials on sugar beet 1962–1965. I.C.I. Farming Service.

60. COOKE, D. A. and DRAYCOTT, A. P. (1971). The effects of soil fumigation and nitrogen fertilisers on nematodes and sugar beet in sandy soils, *Ann. appl. Biol.* **69**, 253–64.

61. COOKE, G. W. (1949). Placement of fertilisers for row crops, *J. agric. Sci., Camb.* **39**, 359–73.

62. COOKE, G. W. (1951). Placement of fertilisers for sugar beet, *J. agric. Sci., Camb.* **41**, 174–8.

63. COOKE, G. W. (1953). The correlation of easily soluble phosphorus in soils with responses of crops to dressings of phosphate fertilisers, *J. Sci. Fd. Agric.* **4**, 353–63.

64. COOKE, G. W. (1954). Recent developments in the use of fertilisers, *Agric. Prog.* **29**, 110.

65. COOKE, G. W. (1964). Soils and fertilisers, *Jl. R. agric. Soc.* **125**, 142–68.

66. COOKE, G. W. (1967). 'The Control of Soil Fertility.' London, Crosby Lockwood and Son Ltd.

67. COOKE, G. W. and WILLIAMS, R. J. B. (1972). Problems with cultivations and soil structure at Saxmundham, *Rep. Rothamsted Exp. Stn. for 1971*, Part 2. In the press.

68. COOMBE, A. and DUNDAS, J. (1960). Post-emergence nitrate of soda sprays for combined nitrogenous fertilisation and weed control in sugar beet, *Proc. Br. Weed Control Conf.* **1**, 47–52.

69. COPE, F. and HUNTER, I. G. (1967). Interactions between nitrogen and phosphate in agriculture. 'Phosphorus in Agriculture', Bull. Doc. No. 46.

70. CROHAIN, A. and RIXHON, L. (1967). [Practical fertilising value of sugar-beet leaves and crowns], *Bull. Rech. agron. Gembloux* **2**, 397–428.

71. CROWTHER, E. M. and YATES, F. (1941). Fertiliser policy in war-time: the fertiliser requirements of arable crops, *Emp. J. exp. Agric.* **9**, 77–97.

72. CROWTHER, E. M. (1947). The use of salt for sugar beet, *Br. Sug. Beet Rev.* **16**, 19–22.

73. DADD, C. V. and BULLEN, E. R. (1958). Manurial requirements of light fen peat soils II, *Expl. Husb.* **3**, 1–17.

74. DAVIS, J. F., SUNDQUIST, W. B. and FRAKES, M. G. (1959). The effect of fertilisers on sugar beets including an economic optima study of response, *J. Am. Soc. Sug. Beet Technol.* **10**, 424–34.

75. DAVIS, J. F., NICHOL, G. and THURLOW, D. (1962). The interaction of rates of phosphate application with fertiliser placement and fertiliser applied at planting time on the chemical composition of sugar beet tissue, yield, percent sucrose and apparent purity of sugar beet roots, *J. Am. Soc. Sug. Beet Technol.* **12**, 259–67.

76. DEVINE, J. R. (1962). The comparative agronomic value of fertilisers in solid and liquid form. International Superphosphate Manufacturers Association, *Extr. Bull. Docum.* **33**, 17–28.

77. DEVINE, J. R. and HOLMES, M. R. J. (1963). Field experiments on the value of urea as a fertiliser for barley, sugar beet, potatoes, winter wheat and grassland in Great Britain, *J. agric. Sci., Camb.* **61**, 391–6.

78. DESPREZ, V. (1963). Influence of the spraying of urea on the yield and content of sugar beet, *Proc. Int. Inst. Sug. Beet Res. XXVI Congr., Brussels.*

79. DOXTATOR, C. W. and CARLTON, F. R. (1950). Sodium and potassium content of sugar beet varieties in some Western beet growing areas, *Proc. Am. Soc. Sug. Beet Technol.* **6**, 144–151.

80. DRAYCOTT, A. P. and COOKE, G. W. (1966). The effects of potassium fertilisers on quality of sugar beet, *Potass. Symp.* 1966, 131–5.

81. DRAYCOTT, A. P., HODGSON, D. R. and HOLLIDAY, R. (1967). Recent research on the value of fertilisers in solution, *Agric. Prog.* **42**, 68–81.

82. DRAYCOTT, A. P. (1968). Field comparisons between compound and straight fertilisers for sugar beet, *Expl. Husb.* **17**, 8–11.

83. DRAYCOTT, A. P. and DURRANT, M. J. (1969). The effects of magnesium fertilisers on yield and chemical composition of sugar beet, *J. agric. Sci., Camb.* **72**, 319–24.

84. DRAYCOTT, A. P. (1969). The effect of farmyard manure on the fertiliser requirements of sugar beet, *J. agric. Sci., Camb.* **73**, 119–24.

85. DRAYCOTT, A. P. and DURRANT, M. J. (1969). Magnesium fertilisers for sugar beet (Part I), *Br. Sug. Beet Rev.* **37**, 175–9.

86. DRAYCOTT, A. P. and DURRANT, M. J. (1970). Magnesium fertilisers for sugar beet (Part II), *Br. Sug. Beet. Rev.* **38**, 175–80.

87. DRAYCOTT, A. P. and HOLLIDAY, R. (1970). Comparisons of liquid and solid fertilisers and anhydrous ammonia for sugar beet, *J. agric. Sci., Camb.* **74**, 139–45.

88. DRAYCOTT, A. P. and LAST, P. J. (1970). Effect of previous cropping and manuring on the nitrogen fertiliser needed by sugar beet, *J. agric. Sci., Camb.* **74**, 147–52.

89. DRAYCOTT, A. P., MARSH, J. A. P. and TINKER, P. B. H. (1970). Sodium and potassium relationships in sugar beet, *J. agric. Sci., Camb.* **74**, 568–73.

90. DRAYCOTT, A. P. and DURRANT, M. J. (1970). The relationship between exchangeable soil magnesium and response by sugar beet to magnesium sulphate, *J. agric. Sci., Camb.* **75**, 137–43.

91. DRAYCOTT, A. P., HULL, R., MESSEM, A. B. and WEBB, D. J. (1970). Effects of soil compaction on yield and fertiliser requirement of sugar beet, *J. agric. Sci., Camb.* **75**, 533–7.

92. DRAYCOTT, A. P. (1970). The growth of sugar beet roots in relation to moisture extraction from the soil profile, *J. int. Inst. Sugar Beet Res.* **5**, 65–70.

93. DRAYCOTT, A. P. (1970). Sugar beet manuring—magnesium, *Rep. Rothamsted Exp. Stn. for* 1969, 327.

94. DRAYCOTT, A. P. (1971). Fertiliser requirements of sugar beet on peaty mineral and organic soils, *Expl. Husb.* **20**, 64–8.

95. DRAYCOTT, A. P. (1971). Anhydrous ammonia compared with solid fertiliser for sugar beet, Anhydrous Ammonia Symposium, 1970, pp. 69–71. I.P.C. Business Press Ltd.

96. DRAYCOTT, A. P. and LAST, P. J. (1971). Some effects of partial sterilisation on mineral nitrogen in a light soil, *J. Soil Sci.* **22**, 152–7.

97. DRAYCOTT, A. P. and WEBB, D. J. (1971). Effects of nitrogen fertiliser, plant population and irrigation on sugar beet. Part I, *J. agric. Sci., Camb.* **76**, 261–7.

98. DRAYCOTT, A. P. and DURRANT, M. J. (1971). Effects of nitrogen fertiliser, plant population and irrigation on sugar beet. Part II, *J. agric. Sci., Camb.* **76**, 269–75.

99. DRAYCOTT, A. P. and DURRANT, M. J. (1971). Effects of nitrogen fertiliser, plant population and irrigation on sugar beet. Part III, *J. agric. Sci., Camb.* **76**, 277–82.

100. DRAYCOTT, A. P. DURRANT, M. J. and BOYD, D. A. (1971). The relationship between soil phosphorus and response by sugar beet to phosphate fertiliser on mineral soils, *J. agric. Sci., Camb.* **77**, 117–21.

101. DRAYCOTT, A. P. and DURRANT, M. J. (1971). Prediction of the fertiliser needs of sugar beet grown on fen peat soils, *J. Sci. Fd. Agric.* **22**, 295–7.

102. DRAYCOTT, A. P. and FARLEY, R. F. (1971). The effect of sodium and magnesium fertilisers and irrigation on growth, composition and yield of sugar beet, *J. Sci. Fd. Agric.* **22**, 559–63.

103. DRAYCOTT, A. P. DURRANT, M. J. and LAST, P. J. (1971). Effects of cultural practices and fertilisers on sugar beet quality, *J. int. Inst. Sugar Beet Res.* **5**, 169–85.

104. DRAYCOTT, A. P., DURRANT, M. J. and WEBB, D. J. (1972). Long-term effects of fertilisers at Broom's Barn, 1965–70, *Rep. Rothamsted Exp. Stn. for* 1971, Part 2, 155–64.

105. DRAYCOTT, A. P., DURRANT, M. J., HULL, R. and WEBB, D. J. (1972). Yields of sugar beet and barley in contrasting crop rotations at Broom's Barn, 1965–70, *Rep. Rothamsted Exp. Stn. for* 1971, Part 2, 149–54.

106. DUBOURG, J., SAUNIER, R. and DEVILLERS, P. (1957). Influence des engrais azotés sur la teneur des Betteraves en constituants azotés et particulièrement en acid glutamique, *Industr. Agric.* **74**, 883–88.

107. DUNNING, R. A. and WINDER, G. H. (1967). Nematicide trials, *Rep. Rothamsted Exp. Stn. for* 1966, 284–6.

108. DUNNING, R. A. and COOKE, D. A. (1967). Docking disorder, *Br. Sug. Beet Rev.* **36**, 23–9.

109. DUNNING, R. A. and WINDER, G. H. (1972). Some effects, especially on yield, of artificially defoliating sugar beet, *Ann. appl. Biol.* In the press.

110. DURRANT, M. J. and DRAYCOTT, A. P. (1971). Uptake of magnesium and other fertiliser elements by sugar beet grown on sandy soils, *J. agric. Sci., Camb.* **77**, 61–8.

111. DURRANT, M. J., DRAYCOTT, A. P. and BOYD, D. A. (1972). The response by sugar beet to potassium and sodium fertilisers, Parts I and II. In preparation.

112. DYKE, G. V. (1965). Green manuring for sugar beet, *Br. Sug. Beet Rev.* **34**, 94–8.

113. EATON, F. M. (1944). Deficiency, toxicity and accumulation of boron in plants, *J. agric. Res.* **69**, 237–77.

114. EDINBURGH AND EAST OF SCOTLAND COLLEGE OF AGRICULTURE (1957). Experiments on level and time of application of nitrogenous fertiliser for sugar beet 1954–56, *Rural Advisory Leaflet* **40**, 1–3.

115. ELLERTON, S. (1947). An experiment to show the effect of nitrogenous fertiliser and of 'topping' the stem on the yield and quality of sugar beet seed, *Proc. Int. Inst. Sug. Beet Res. X Congr. Brussels.*

116. ELLERTON, S. (1951). Results of a sugar beet seed yield trial conducted at Woodham Mortimer, Maldon, Essex, in 1950. Personal communication.

117. ELLERTON, S. (1958). Experiments in sugar beet seed growing technique. Personal communication.

118. EL-SHEIKH, A. M., ULRICH, A. and BROYER, T. C. (1967). Sodium and rubidium as possible nutrients for sugar beet plants, *Pl. Physiol., Lancaster.* **42**, 1202–8.

119. FIELDLER, J. (1972). Private communication.

120. FINKNER, R. E. and BAUSERMAN, H. M. (1956). Breeding of sugar beets with reference to sodium, sucrose and raffinose content, *J. Am. Soc. Sug. Beet Technol.* **9**, 170–7.

121. FINKNER, R. E., OGDEN, D. B., HANZAS, P. C. and OLSON, R. F. (1958). The effect of fertiliser treatment on the calcium, sodium, potassium, raffinose, galactinol, nine amino acids, and total amino acid content of three varieties of sugar beets grown in the Red Valley of Minnesota, *J. Am. Soc. Sug. Beet Technol.* **10**, 272–80.

122. FORD, E. M. (1968). Studies in the nutrition of apple root stocks. V. The development of magnesium deficiency symptoms in relation to magnesium supply, *Ann. Bot.* **32**, 45–56.

123. FUEHRING, H. D., HASHIMI, M. A., HADDAD, K. S., HUSSIENI, K. K. and MAKHDOOM, M. U. (1969). Nutrient interacting effects on sucrose yield of sugar beets (*Beta vulgaris*) on a calcareous soil, *Proc. Soil Sci. Soc. Am.* **33**, 718–21.

124. GALLAGHER, P. A. (1967). Fertiliser experiments on sugar beet in Eire, *Inf. Nitr. Corp. Chile* **98**.

125. GARBOUCHEV, I. P. (1966). Changes occurring during a year in the soluble phosphorus and potassium in soil under crops in rotation experiments at Rothamsted, Woburn and Saxmundham, *J. agric. Sci., Camb.* **66**, 399–412.

126. GARNER, H. V. (1950). Sugar beet irrigation, *Br. Sug. Beet Rev.* **18**, 145–50.

127. GARNER, H. V. (1966). Experiments on the direct, cumulative and residual effects of town refuse manures and sewage sludge at Rothamsted and other centres 1940–1947, *J. agric. Sci., Camb.* **67**, 223–34.

128. GARNER, H. V. (1968). Field experiments on carrots at Rothamsted, Woburn and Tunstall (Suffolk), *Exp. Hort.* **18**, 69–76.

129. GASCHO, G., DAVIS, J. F., FOGG, R. A. and FRAKES, M. G. (1969). The effects of potassium carriers and levels of potassium and nitrogen fertilisation on the yield and quality of sugar beets, *J. Am. Soc. Sug. Beet Technol.* **15**, 298–305.

130. GASKILL, J. O. (1950). Progress report on the effects of nutrition, bruising and washing upon rotting of stored sugar beets, *Proc. Am. Soc. Sug. Beet Technol.* **6**, 680–5.

131. GERICKE, S. (1966). Phosphate fertilising and quality of the sugar beet crop, *Zucker* **24**, 663–7.

132. GOODMAN, P. J. (1963). Some effects of different soils on composition and growth of sugar beet, *J. Sci. Fd. Agric.* **14**, 196–203.

133. GOODMAN, P. J. (1966). Effect of varying plant populations on growth and yield of sugar beet, *Agric. Prog.* **41**, 89–107.

134. GOODMAN, P. J. (1968). Physiological analysis of the effects of different soils on sugar beet crops in different years, *Jnl. appl. Ecol.* **5**, 339–57.

135. GRAF, A. (1972). Private communication.

136. GREENHAM, D. W. P. (1968). Movement of potassium and magnesium in fruit soils. Paper presented to the N.A.A.S. 'Open' conference of Soil Scientists.

137. GREGG, C. M. and HARRISON, C. M. (1950). A study of the effects of some different sods and fertilisers on sugar beet yields, *Proc. Am. Soc. Sug. Beet Technol.* **6**, 306–10.

138. GRÖNEVIK, G. (1972). Private communication.

139. GÜRAY, R. (1972). Private communication.

140. GUREVICH, S. M. and BORONINA, I. I. (1964). [Uptake and removal of nutrients by sugar beet in relation to the level of nutrition], *Agrokhimiya* **10**, 73–81.

141. GUTSTEIN, Y. (1970). Response of winter-sown sugar beet cultivars to manure and nitrogen rates and time of application, *Israel J. agric. Res.* **20**, 41–5.

142. HADDOCK, J. L. and KELLEY, O. J. (1948). Inter-relations of moisture, spacing and fertility to sugar beet production, *Proc. Am. Soc. Sug. Beet Technol.* **5**, 378–96.

143. HADDOCK, J. L. (1949). The influence of plant population, soil moisture, and nitrogen fertilisation on the sugar content and yield of sugar beets, *Agron. J.* **41**, 79–84.

144. HADDOCK, J. L., LINTON, D. C. and HURST, R. C. (1956). Nitrogen constituents associated with reduction of sucrose percentage and purity of sugar beets, *J. Am. Soc. Sug. Beet Technol.* **9**, 110–17.

145. HADDOCK, J. L., SMITH, P. B., DOWNIE, A. R., ALEXANDER, J. T., EASTON, B. E. and JENSEN, V. (1959). The influence of cultural practices on the quality of sugar beets, *J. Am. Soc. Sug. Beet Technol.* **10**, 290–301.

146. HADDOCK, J. L. (1959). Yield, quality and nutrient content of sugar beets as affected by irrigation regime and fertilisers, *J. Am. Soc. Sug. Beet Technol.* **10**, 344–55.

147. HADDOCK, J. L. and STUART, D. M. (1970). Nutritional conditions in sugar beet fields of Western United States and chemical composition of leaf and petiole tissue, including minor elements, *J. Am. Soc. Sug. Beet Technol.* **15**, 684–702.

148. HAGEN, R. J. (1968). Beet without ploughing, *Br. Sug. Beet Rev.* **36**, 180–9.

149. HALE, J. B. (1945). Deficiency diseases of the sugar beet. Agric. Res. Coun. Unpublished report, duplicated for private circulation A.R.C. 7828, p. 8.

150. HALE, J. B., WATSON, M. A. and HULL, R. (1946). Some causes of chlorosis and necrosis of sugar beet foliage, *Ann. appl. Biol.* **33**, 13–28.

151. HALE, V. Q. and MILLER, R. J. (1966). Relationships between NO_3^-–N in petioles during the growing season and yield components of sugar beets (*Beta vulgaris*), *Agron J.* **58**, 567–9.

152. HALL, A. D. (1902). The continuous growth of mangolds for twenty-seven years on the same land, Barnfield, Rothamsted, *Jl. R. agric. Soc.* **63**, 27–59.

153. HAMENCE, J. H. and ORAM, P. A. (1964). Effects of soil and foliar applications of sodium borate to sugar beet, *J. Sci. Fd. Agric.* **15**, 565–79.

154. HANLEY, F. and MANN, J. C. (1936). The control of heart rot in sugar beet, *J. Min. Agric.* **43**, 15–23.

155. HANLEY, F. (1951). Economics of factory carbonate of lime, *Br. Sug. Beet Rev.* **19**, 145–7.

156. HANLEY, F. and RIDGMAN, W. J. (1956). Manurial requirements of light fen peat soils: I. *Expl. Husb.* **1**, 1–9.

157. HANLEY, F. and RIDGMAN, W. J. (1963). An investigation into the long-term effects of combine drilling on the yield of arable crops, *Expl. Husb.* **9**, 19–27.

158. HARMER, P. M., BENNE, E. J., LAUGHLIN, W. M. and KEY, C. (1953). Factors affecting crop response to sodium applied as common salt on Michigan muck soil, *Soil Sci.* **76**, 1–17.

159. HARRIS, P. M. (1969). A study of the interaction between method of establishing and method of harvesting the sugar beet crop, *J. int. Inst. Sugar Beet Res.* **4**, 84–103.

160. HARRIS, P. M. (1970). The interaction between plant density and irrigation in the sugar beet crop, *Proc. Int. Inst. Sug. Beet Res. XXXIII Congr., Brussels.*

161. HARROD, M. F. and CALDWELL, T. H. (1967). The magnesium manuring of sugar beet on light sandy soils of East Anglia, *Tech. Bull. Minist. Agric. Fish. Fd.* **14**, 127–42, London, HMSO.

162. HARVEY, P. N. (1959). The disposal of cereal straw, *Jl. R. agric. Soc.* **120**, 55–63.

163. HEALD, W. R., MOODIE, C. D. and LEAMER, R. W. (1950). Leaching and pre-emergence irrigation for sugar beets on saline soils, *Bull. Wash. agric. Exp. Stn.* 519.

164. HEATHCOTE, G. D. (1970). Effect of plant spacing and time of sowing of sugar beet on aphid infestation and spread of virus yellows, *Pl. Path.* **19**, 32–9.

165. HEATHCOTE, G. D. (1972). Influence of cultural factors on incidence of aphids and yellows in beet, *J. int. Inst. Sugar Beet Res.* In the press.

166. HEISTERMANN, P. (1968). [Yield and quality of sugar beet and potatoes as affected by fertiliser N:K ratio], *Zesz. probl. Postep. Nauk roln.* **84**, 273–88.

167. HEMINGWAY, R. G. (1961). The mineral composition of farmyard manure, *Emp. J. exp. agric.* **29**, 14–18.

168. HENDERSON, D. W., HILLS, F. J., LOOMIS, R. S. and NOURSE, E. F. (1968). Soil moisture conditions, nutrient uptake and growth of sugar beets as related to method of irrigation of an organic soil, *J. Am. Soc. Sugar Beet Technol.* **15**, 35–48.

169. HENKENS, C. H. and JONGMAN, E. (1965). The movement of manganese in the plant and the practical consequences, *Neth. J. agric. Sci.* **13**, 392–407.

170. HENKENS, C. H. and SMILDE, K. W. (1966). Evaluation of glassy frits as micronutrient fertilisers. I. Copper and molybdenum frits, *Neth. J. agric. Sci.* **14**, 165–77.

171. HENKENS, C. H. and SMILDE, K. W. (1967). Evaluation of glassy frits as micronutrient fertilisers. II. Manganese frits, *Neth. J. agric. Sci.* **15**, 21–30.

172. HERA, C., DAVIDESCU, D. and VINES, I. (1961). [The effect of ammonia liquor and anhydrous ammonia on the yields of sugar beet, silage maize and oats], *Lucr. stiint. Inst. agron. Nicolae Balcescu* **5A**, 163–9.

173. HERNANDO, V., JIMENO, L. and GUERRA, A. (1961). [Study of the effect of the time of application of the nitrate on the production of sugar beet], *An. Edafol. Agrobiol.* **20**, 9–10.

174. HERRON, G. M., GRIMES, D. W. and FINKNER, R. E. (1964). Effect of plant spacing and fertiliser on yield, purity, chemical constituents and evapotranspiration of sugar beets in Kansas. Parts I and II, *J. Am. Soc. Sug. Beet Technol.* **12**, 699–714.

175. HEWITT, E. J. (1948). Relation of manganese and some other metals to the iron status of plants, *Nature, Lond.* **161**, 489–90.

176. HEWITT, E. J. (1953). Metal inter-relations in plants. Part I. Effects of some metal toxicities on sugar beet, tomato, oat, potato, and marrowstem kale grown in sand culture, *J. exp. Bot.* **4**, 59–64.

177. HILLS, F. J., BURTCH, L. M., HOLMBERG, D. M. and ULRICH, A. (1954). Response of yield-type versus sugar-type sugar beet varieties to soil nitrogen levels and time of harvest, *Proc. Am. Soc. Sug. Beet Technol.* **8** (1), 64–70.

178. HILLS, F. J., FERRY, G. V., ULRICH, A. and LOOMIS, R. S. (1963). Marginal nitrogen deficiency of sugar beets and the problems of diagnosis, *J. Am. Soc. Sug. Beet Technol.* **12**, 476–84.

179. HILLS, F. J., SAILSBERY, R. L., ULRICH, A. and SIPITANOS, K. M. (1970). Effect of phosphorus on nitrate in sugar beet (*Beta vulgaris*), *Agron. J.* **62**, 91–92.

180. HILLS, F. J. and ULRICH, A. (1971). Nitrogen nutrition, *in* 'Advances in Sugar Beet Production: Principles and Practices,' pp. 111–36. Ames, Iowa, Iowa State University Press.

181. HOLLIDAY, R., HARRIS, P. M. and BABA, M. R. (1965). Investigations into the mode of action of farmyard manure. I. The influence of soil moisture conditions on the response of maincrop potatoes to farmyard manure, *J. agric. Sci., Camb.* **64**, 161–6.

182. HOLMES, J. C., GILL, W. D., RODGER, J. B. A., WHITE, G. R. and LAWLEY, D. N. (1961). Experiments with salt and potash on sugar beet in South-East Scotland, *Expl. Husb.* **6**, 1–7.

183. HOLMES, J. C. and ADAMS, S. N. (1966). The effect of sowing date, harvest date and fertiliser rate on sugar beet, *Expl. Husb.* **14**, 65–74.

183A. HOOPER, L. J. (1970). The basis of current fertiliser recommendations in England and Wales, *Proc. Fertil. Soc.* 118.

184. HOYT, P. B. (1968). The effect of soil conditioners on the growth of sugar beet in a sandy loam soil. *Expl. Husb.* **16**, 70–2.

185. HULL, R. and WILSON, A. R. (1946). Distribution of violet root rot (*Helicobasidium purpureum Pat.*) of sugar beet and preliminary experiments on factors affecting the disease, *Ann. appl. Biol.* **33**, 420–33.

186. HULL, R. and WATSON, M. (1947). Factors affecting the loss of yield of sugar beet caused by beet yellows virus. II. Nutrition and variety, *J. agric. Sci., Camb.* **37**, 301–10.

187. HULL, R. (1960) (2nd Edn). Sugar beet diseases, *Tech. Bull. Minist. Agric. Fish. Fd.* 142, London, HMSO.

188. HULL, R. (1965). Control of sugar-beet yellows. Symposium on some approaches towards integrated control of British insect pests, *Ann. appl. Biol.* **56**, 345–7.

189. HULL, R. and WEBB, D. J. (1967). The effect of subsoiling and different levels of manuring on yields of cereals, lucerne and sugar beet, *J. agric. Sci., Camb.* **69**, 183–7.

189A. HULL, R. and WEBB, D. J. (1970). The effect of sowing date and harvesting date on the yield of sugar beet, *J. agric. Sci., Camb.* **75**, 223–9.

190. HULL, R. (1970). Personal communication.

191. INTERNATIONAL POTASH INSTITUTE (1955). The sugar beet and its manuring, Berne.

192. JACOB, A. (1958). 'Magnesium, the Fifth Major Plant Nutrient.' London; Staples Press.

193. JAMES, D. W., LEGGETT, G. E. and DOW, A. I. (1967). Phosphorus fertility relationships of central Washington irrigated soils, with special emphasis on exposed calcareous subsoils, *Bull. Wash. agric. Exp. Stn. No.* 688, 1–18.

194. JAMESON, H. R. (1959). Liquid nitrogenous fertilisers, *J. agric. Sci., Camb.* **53**, 333–8.

195. JOHNSTON, A. E., WARREN, R. G. and PENNY, A. (1970). The value of residues from long-period manuring at Rothamsted and Woburn. V. The value to arable crops of residues accumulated from potassium fertilisers, *Rep. Rothamsted Exp. Stn. for* 1969, Part 2, 69–90.

196. JÓNSSON, L. (1969). [Effect of phosphate fertilising on the efficiency of nitrogen for sugar beet], *LantbrHögsk. Meddn* **120A**, 18.

197. JORRITSMA, J. (1956). [The manuring of sugar beet. I. N–K–Mg trial fields 1946–51 inclusive], *Meded. Inst. rat. SuikProd* **26**, 227.

198. JORRITSMA, J. (1961). The fertilising of sugar beet. II. Nitrogen fertilisers, *Meded. Inst. rat. SuikProd* **2**, 53–157.

199. JORRITSMA, J. (1967). Nitrogen manuring experiments on sugar beet in the Netherlands, *J. int. Inst. Sugar Beet Res.* **2**, 69–85.

200. JORRITSMA, J. (1972). Private communication.

201. KANWAR, J. S. (1969). Sugar beet for saline and alkali soils in Northern India, *Indian Fmg.* **19**, 5–6.

202. KELLEY, J. D. and ULRICH, A. (1966). Distribution of nitrate nitrogen in the blades and petioles of sugar beets grown at deficient and sufficient levels of nitrogen, *J. Am. Soc. Sug. Beet Technol.* **14**, 106–16.

203. KNOWLES, F., WATKIN, J. E. and HENDRY, F. W. F. (1934). A chemical study of sugar beet during the first growth year, *J. agric. Sci., Camb.* **24**, 368–78.

204. KOTILA, J. E. and COONS, G. H. (1935). Boron deficiency disease of beets, *Facts Sug.* **30**, No. 10.

205. KOZERA, H. and LACHOWSKI, J. (1959). [Some effects of preharvest foliage sprays on sugar beet plants], *Bull. Inst. How. Aklin. Ros.* 73–84.

206. KRÜGEL, C., DREYSPRING, C. and LOTTHAMMER, R. (1938). Leaching experiments with borates, *Superphosphate* **8 and 9**, 141–50, 161–6.

207. KUBOTA, T. and WILLIAMS, R. J. B. (1967). The effects of changes in soil compaction and porosity on germination, establishment and yield of barley and globe beet, *J. agric. Sci., Camb.* **68**, 227–33.

208. KUPERS, L. J. P. and ELLEN, J. (1970). Experience with minimum tillage and nitrogen fertilisation, *Neth. J. agric. Sci.* **18**, 270–6.

209. LACHOWSKI, J. (1959). [Comparison of pulverised granulated superphosphate in fertilising sugar beets], *Roczn. Nauk rol.* **80**, 134–47.

210. LACHOWSKI, J. (1960). [The effect of different rates of zinc sulphate on sugar-beet yields in Poland], *Roczn. Nauk rol.* **83A**, 149–65.
211. LACHOWSKI, J. (1961). [The effect of boron, manganese and copper on the production value of sugar beet in Poland], *Roczn. Nauk rol.* **84A**, 63–88.
212. LACHOWSKI, J. and WESOLOWSKI, F. (1964). [The influence of iron sulphate on the yield of sugar beets in Poland], *Roczn. Nauk rol.* **89A**, 547–64.
213. LACHOWSKI, J. (1966). [The effect of magnesium upon the growth and crop of root plants of sugar beets, and the progeny obtained after them], *Hodowla Rosl. Aklim. Nasienn.* **10**, 37–59.
214. LACHOWSKI, J. (1968). [A study of the growth and of the nutritive element content in some varieties of the sugar beet], *Buil. Inst. How. Aklin. Ros.* **5–6**, 127–33.
215. LARMER, F. G. (1937). Keeping quality of sugar beets as influenced by growth and nutritional factors, *J. agric. Res.* **54**, 185–98.
216. LARSEN, S. (1964). The effect of phosphate application on manganese content of plants grown on neutral and alkaline soils, *Pl. Soil* **21**, 37–42.
217. LAST, P. J. and TINKER, P. B. H. (1968). Nitrate nitrogen in leaves and petioles of sugar beet in relation to yield of sugar and juice purity, *J. agric. Sci., Camb.* **71**, 383–92.
218. LAST, P. J. and DRAYCOTT, A. P. (1971). Predicting the amount of nitrogen fertiliser needed for sugar beet by soil analysis, *J. Sci. Fd. Agric.* **22**, 215–20.
219. LAST, P. J. and DRAYCOTT, A. P. (1972). Top-dressing of nitrogen for sugar beet, *Expl. Husb.* In the press.
220. LEACH, L. D. and DAVEY, A. E. (1942). Reducing southern sclerotium rot of sugar beets with nitrogenous fertilisers, *Agric. Res. Wash.* **64**, 1–18.
221. LE DOCTE, A. (1927). Commercial determination of sugar in the beet root using the Sachs–Le Docte process, *Int. Sug. J.* **29**, 488–92.
222. LEE, J. and GALLAGHER, P. A. (1970). Influence of the soil factor on sugar yields and responses to nutrients, *Ir. J. agric. Res.* **9**, 143–7.
223. LEHR, J. J. and BUSSINK, A. TH. (1962). Nitrogen and potash fertilisation of sugar beet on sandy soils. Experiments in Holland 1958–61, *Inf. Nitr. Corp. Chile* 2–11.
224. LEONARD, C. D. and BEAR, F. E. (1950). Sodium as a fertiliser for New Jersey soils, *Bull. New Jers. agric. Exp. Stn.* **752**, 24.
225. LEWIS, A. H. (1941). The placement of fertilisers. I. Root crops, *J. agric. Sci., Camb.* **31**, 295–307.
226. LONGDEN, P. C. (1970). Manuring the beet seed crop growth in England, *N.A.A.S. q. Rev.* **87**, 112–18.
227. LONGDEN, P. C. (1972). Effects of some soil factors on sugar-beet seedling emergence. In preparation.
228. LOOMIS, R. S. and NEVINS, D. J. (1963). Interrupted nitrogen nutrition effects on growth, sucrose accumulation and foliar development of the sugar beet plant, *J. Am. Soc. Sug. Beet Technol.* **12**, 309–22.
229. LOOMIS, R. S. and WORKER, G. F. JR. (1963). Responses of the sugar beet to low soil moisture at two levels of nitrogen nutrition, *Agron. J.* **55**, 509–15.
230. LOOMIS, R. S. and WORKER, G. F. JR. (1964). Restitution of growth in nitrogen deficient sugar beet plants, *J. Am. Soc. Sug. Beet Technol.* **12**, 657–65.
231. LÜDECKE, H. VON and NITZSCHE, M. (1967). Influence of excessive amounts of mineral fertilisers on yield and quality of sugar beets, *Zucker* **20**, 461–6, 483–7.
232. LUGG, G. W. (1971). Economic considerations, Anhydrous Ammonia Symposium 1970, pp. 122–3. I.P.C. Business Press Ltd.

233. McALLISTER, J. S. V. and RUTHERFORD, A. A. (1967). Manurial experiments with sugar beet, 1963–65, *Rec. agric. Res. (Nth. Ireld.)* **16**, 113–22.
234. McDONNELL, P. M., GALLAGHER, P. A., KEARNEY, P. and CARROLL, P. (1966). Fertiliser use and sugar beet quality in Ireland, *Potass. Symp.* 1966, 107–26.
235. McENROE, P. and COULTER, B. (1964). Effect of soil pH on sugar content and yield of sugar beet, *Ir. J. agric. Res.* **3**, 63–9.
236. MACKENZIE, H. (1967). Factory waste lime experience in Lincolnshire, *Br. Sug. Beet Rev.* **35**, 177–8.
237. McMILLAN, J. A. and HANLEY, F. (1936). The effect of sowing fertilisers in contact with the seed of barley and of sugar beet, *Agriculture, Lond.* **42**, 1205–11.
238. MARKOVIC, N. and STOJANOVIC, Z. (1966). [Fixing of the most appropriate ratio of N P K fertilisers for sugar beet and sunflower on chernozem and forest soil], *Zemlj. Biljka* **15**, 339–52.
239. MANN, H. H. and BOYD, D. A. (1958). Some results of an experiment to compare ley and arable rotations at Woburn, *J. agric. Sci., Camb.* **50**, 297–306.
240. MANN, H. H. and PATTERSON, H. D. (1963). The Woburn market-garden experiments, Summary 1944–60, *Rep. Rothamsted Exp. Stn. for 1962*, 186–93.
241. MANN, J. and BARNES, T. W. (1945). Manuring for the production of sugar beet seed, *Agriculture, Lond.* **52**, 400–4.
242. MARTENS, M. and OLDFIELD, J. F. T. (1970). Storage of sugar beet in Europe: Report of an I.I.R.B. enquiry, *J. int. Inst. Sugar Beet Res.* **5**, 102–27.
243. MATTINGLY, G. E. G., JOHNSTON, A. E. and CHATER, M. (1970). The residual value of farmyard manure and superphosphate in the Saxmundham Rotation II experiment, 1899–1968, *Rep. Rothamsted Exp. Stn. for 1969*, Part 2, 91–112.
244. MEYER, H. (1972). Private communication.
245. MICZYNSKI, J. and SIWICKI, S. (1954). [Intercrop manuring in cultivation of sugar beet], *Roczn. Nauk rol.* **70A**, 251–81.
246. MICZYNSKI, J. and SIWICKI, S. (1959). [Investigations on the influence of green manure on sugar beets], *Buil. Inst. How. Aklin. Ros.* 39–60.
247. MILES, R. O. (1947). The placement of fertilisers, *Bull. Jealotts Hill Res. Stn.* **4**.
248. MILFORD, G. F. and WATSON, D. J. (1971). The effect of nitrogen on the growth and sugar content of sugar beet, *Ann. Bot.* **35**, 287–300.
249. MORLEY DAVIES, W. (1939). Acidity and manganese deficiency problems in connexion with sugar beet growing, *Ann. appl. Biol.* **26**, 285–392.
250. MORLEY DAVIES, W. (1943). Minor elements and crop failures, *Agric. Prog.* **18**, 28–33.
251. MURPHY, L. S. and SMITH, G. E. (1967). Nitrate accumulation in forage crops, *Agron. J.* **59**, 171–4.
252. NAGARAJAH, S. and ULRICH, A. (1966). Iron nutrition of the sugar beet plant in relation to growth, mineral balance, and riboflavin formation, *Soil Sci.* **102**, 399–407.
253. NELSON, J. M. (1969). Effect of row width, plant spacing, nitrogen rate and time of harvest on yield and sucrose content of sugar beets, *J. Am. Soc. Sug. Beet Technol.* **15**, 509–16.
254. NELSON, R. T. (1950). Fertilisers ploughed under versus soil application at or after planting, *Proc. Am. Soc. Sug. Beet Technol.* **6**, 436–9.
255. NEVINS, D. J. and LOOMIS, R. S. (1970). Nitrogen nutrition and photosynthesis in sugar beet (*Beta vulgaris L.*), *Crop Sci.* **10**, 21–5.

256. NITRATE CORP. OF CHILE (1963). Sugar beet in Germany, *Inf. Nitr. Corp. Chile,* June 1963.
257. NOWICKI, A. (1969). [The effect of molybdenum upon yield, health and processing quality of sugar beet], *Roczn. Nauk rol.* **95,** 55–74.
258. ØDELEIN, M. (1963). Long-term field experiments with small applications of boron, *Soil Sci.* **95,** 60–2.
259. OGDEN, D. B., FINKNER, R. F., OLSON, R. F. and HANZAS, P. C. (1958). The effect of fertiliser treatment upon three different varieties in the Red River Valley of Minnesota for: 1. Stand, yield, sugar purity and non-sugars, *J. Am. Soc. Sug. Beet Technol.* **10,** 265–71.
260. OIEN, S. (1972). Private communication.
261. OLSEN, S. R., GARDNER, R., SCHMEHL, W. R., WATANABE, F. S. and SCOTT, C. O. (1950). Utilisation of phosphorus from various fertiliser materials by sugar beets in Colorado, *Proc. Am. Soc. Sug. Beet Technol.* **6,** 317–31.
262. OLSEN, S. R., COLE, C. V., WATANABE, F. S. and DEAN, L. A. (1954). Estimation of available phosphorus in soils by extraction with sodium bicarbonate, *U.S. Dep. Agric. Circ.* 939.
263. ONTANON, J. M. (1960). Sugar beet in Spain, *Fertil. Feed Stuffs. J.* **53,** 276–7.
264. OVERPECK, J. C. and ELCOCK, H. A. (1937). Sugar-beet seed production studies in Southern New Mexico, 1931–36, *Bull. New Mex. agric. Exp. Stn.* 252.
265. PATTERSON, H. D. (1960). An experiment on the effects of straw ploughed in or composted on a three-course rotation of crops, *J. agric. Sci., Camb.* **54,** 222–30.
266. PATTERSON, H. D. and WATSON, D. J. (1960). Farmyard manure and its interaction with fertilisers, *Rep. Rothamsted Exp. Stn. for* 1959, 164–8.
267. PENDLETON, R. A., FINNELL, H. E. and REIMER, F. C. (1950). Sugar beet seed production in Oregon, *Bull. Oregon St. Coll. Agric. Exp. Stn.* 437.
268. PENDLETON, R. A. (1950). Soil compaction and tillage operation effects on sugar beet root distribution and seed yields, *Proc. Am. Soc. Sug. Beet Technol.* **6,** 278–85.
269. PENDLETON, R. A. (1954). Forms of nitrogen as related to sugar beet seed production in Oregon, *Proc. Am. Soc. Sug. Beet Technol.* **8,** 140–1.
270. PENDLETON, R. A. (1954). Cultural practices related to yields and germination of sugar beet seed, *Proc. Am. Soc. Sug. Beet Technol.* **8,** 157–60.
271. PENMAN, H. L. (1948). Evaporation in nature, *Prog. Phys.* **11,** 366–88.
272. PENMAN, H. L. (1952). Experiments on the irrigation of sugar beet, *J. agric. Sci., Camb.* **42,** 286–92.
273. PENMAN, H. L. (1962). Woburn irrigation, 1951–59, Parts I, II, III, *J. agric. Sci., Camb.* **58,** 343–8, 349–64, 365–79.
274. PENMAN, H. L. (1970). Woburn irrigation 1960–68, Parts IV, V, VI, *J. agric. Sci., Camb.* **75,** 69–73, 75–88, 89–102.
275. PETERSON, G. A., ANDERSON, F. N. and OLSON, R. A. (1966). A survey of the nutrient status of soils in the North Platte Valley of Nebraska for sugar production, *J. Am. Soc. Sug. Beet Technol.* **14,** 48–60.
276. PETRESCU, O., PICU, I. and ISFAN, D. (1969). [Effect of increasing nitrogen rates on nitrate and free amino acid content of corn and sugar beet leaves in irrigated crops], *Anal. Inst. Cerc. Cer. Pl. teh.* **35,** 445–53.
277. PIZER, N. H. (1954). Organic matter in some eastern counties soils, *N.A.A.S. q. Rev.* **25,** 41–6.
278. PIZER, N. H., CALDWELL, T. H., BURGESS, G. R. and JONES, J. L. O. (1966). Investigations into copper deficiency in crops in East Anglia, *J. agric. Sci., Camb.* **66,** 303–14.

279. PRICE, T. J. A. and HARVEY, P. N. (1961). The effect of irrigating sugar beet on a poor sand soil, *Proc. Int. Inst. Sug. Beet Res. XXIV Congr., Brussels* 317–23.

280. PRICE, T. J. A. and HARVEY, P. N. (1962). Effect of irrigation on sugar beet and potatoes, *Expl. Husb.* **7**, 1–7.

281. PRUMMEL, J. (1957). Fertiliser placement experiments, *Pl. Soil* **8**, 231–53.

282. PULTZ, L. M. (1937). Relation of nitrogen to yield of sugar beet seed and to accompanying changes in the composition of the roots, *J. agric. Res.* **54**, 639–54.

283. RÁB, F. (1968). [Location and translocation of boron in sugar-beet plants], *Acta Univ. Agric. Brno, Fac. Agron.* **16**, 601–7.

284. RASMUSSON, J. and WIKLUND, O. (1960). Characteristics of the technological value of the sugar beet, *Proc. XIth Sess. Comm. Int. Tech. Suc.* 1960, 13–24.

285. RAUHE, K. and HESSE, M. (1959). [The role of organic manuring in increasing the yields of row crops, including maize], *Dtsch. Landw.* **10**, 284–8.

286. RAYNS, F. (1961). A revolution in arable farming. The Lord Hastings Memorial Lecture. Norfolk Agricultural Station.

287. RID, H. (1972). Private communication.

288. ROBERTSON, L. S. (1952). A study of the effects of seven systems of cropping upon yields and soil structure, *Proc. Am. Soc. Sug. Beet Technol.* **7**, 255–64.

289. ROBINS, J. S., NELSON, C. E. and DOMINGO, C. E. (1956). Some effects of excess water application on utilisation of applied nitrogen by sugar beets, *J. Am. Soc. Sug. Beet Technol.* **9**, 180–8.

290. ROMSDAL, S. D. and SCHMEHL, W. R. (1963). The effect of method and rate of phosphate application on yield and quality of sugar beets, *J. Am. Soc. Sug. Beet Technol.* **12**, 603–7.

290A. ROSELL, R. A. and ULRICH, A. (1964). Critical zinc concentrations and leaf minerals of sugar beet plants, *Soil Sci.* **97**, 152–67.

291. ROSCOE, B. (1960). The distribution and condition of soil phosphate under old permanent pasture, *Pl. Soil* **12**, 17–29.

292. ROUSSEL, N., STALLEN, R. VAN and VLASSAK, K. (1966). [Results of two years fertiliser trials with anhydrous ammonia], *J. Int. Inst. Sugar Beet Res.* **2**, 35–52.

293. ROUSSEL, N. (1972). Private communication.

294. ROWE, E. A. (1936). A study of heart-rot of young sugar beet plants grown in culture solutions, *Ann. Bot.* **50**, 735–46.

295. RUSSELL, E. W. (1956). The effects of very deep ploughing and of subsoiling on crop yields, *J. agric. Sci., Camb.* **48**, 129–44.

296. RUSSELL, G. E. (1971). Effects on Myzus persicae (Sulz) and transmission of beet yellows virus of applying certain trace elements to sugar beet, *Ann. appl. Biol.* **68**, 67–70.

297. RYSER, G. K. (1966). A regression study on tare samples of sugar beets in relation to factors influencing productivity and quality, *J. Am. Soc. Sug. Beet Technol.* **13**, 727–47.

298. SALMON, R. C. (1963). Magnesium relationships in soils and plants, *J. Sci. Fd. Agric.* **14**, 605–10.

299. SALTER, P. J. and WILLIAMS, J. B. (1963). The effect of farmyard manure on the moisture characteristic of a sandy loam soil, *J. Soil Sci.* **14**, 73–81.

300. SALTER, P. J., WILLIAMS, J. B. and HARRISON, D. J. (1965). Effects of bulky organic manures on the available water capacity of a fine sandy loam, *Expl. Hort.* **13**, 69–75.

301. SAUCHELLI, V. (1969). 'Trace Elements in Agriculture.' New York, Van Nostrand Reinhold Company.

302. SAWAHATA, H. and TAKASE, N. (1966). [Studies on germination and early growth of sugar beet in Southern Japan. III. Effects of kinds and amounts of nitrogenous fertilisers on germination and early growth of sugar beet], *Res. Meeting Sug. Beet Tech. Co-op. S. Japan* **8**, 26–30.
303. SCHMEHL, W. R., FINKNER, R. and SWINK, J. (1963). Effect of nitrogen fertilisation on yield and quality of sugar beet, *J. Am. Soc. Sug. Beet Technol.* **12**, 538–44.
304. SCHMID, G. (1967). [The effect of liming on crop quality], *Z. Acker-u. PflBau.* **125**, 7–21.
305. SCHREVEN, D. A. VAN (1936). Copper deficiency in sugar beets, *Phytopathology* **26**, 1106–17.
306. SCOTT, R. K. (1967). Sugar beet seed production, *Agric. Prog.* **42**, 112–18.
307. SCOTT, R. K. (1968). Sugar beet seed growing in Europe and North America, *J. int. Inst. Sugar Beet Res.* **3**, 53–84.
308. SCOTT, R. K. (1969). The effect of sowing and harvesting dates, plant population and fertilisers on seed yield and quality of direct-drilled sugar beet seed crops, *J. agric. Sci., Camb.* **73**, 373–85.
309. SHEPPERD, R. W. (1954). The use of nitrogen for sugar beet grown on light land, *Emp. J. exp. Agric.* **22**, 128–32.
310. SHORROCKS, V. M. (1970). Personal communication.
311. SHOTTON, F. E. (1962). Placement of fertiliser for sugar beet, *Expl. Husb.* **7**, 8–16.
312. SIMON, M., ROUSSEL, N. and STALLEN, R. VAN (1966). [Potassium in the fertilising of sugar beet], *Potass. Symp.* 1966, 61–87.
313. SIPITANOS, K. M. and ULRICH, A. (1969). Phosphorus nutrition of sugar beet seedlings, *J. Am. Soc. Sug. Beet Technol.* **15**, 332–46.
314. SIWICKI, S. (1957). The influence of fertiliser applications on three varieties of sugar beets, and profitability of such applications, *Builetyn roslin przemystowych* **7**, 49–75.
315. SMILDE, K. W. (1970). Soil analysis as a basis for boron fertilisation of sugar beets, *Z. PflErnähr. Bodenk.* **125**, 130–43.
316. SMITH, C. H. (1954). The influence of krilium soil conditioner on sugar content of beets, *Proc. Am. Soc. Sug. Beet Technol.* **8**, 143–6.
317. SNEDDON, J. L. (1963). Sugar beet seed production experiments, *J. natn. Inst. agric. Bot.* **9**, 333–45.
318. SNYDER, F. W. (1959). Effect of nitrogen on yield and subsequent germinability of sugar beet seed, *J. Am. Soc. Sug. Beet Technol.* **10**, 439–43.
319. SØRENSEN, C. (1960). The influence of nutrition on the nitrogenous constituents of plants. II. Field experiments with heavy dressings of nitrogen to fodder sugar beets, *Acta Agric. scand.* **10**, 17–32.
320. SØRENSEN, C. (1962). The influence of nutrition on the nitrogenous constituents of plants. III. Nitrate tests and yield structure of fodder sugar beet leaves, *Acta Agric. scand.* **12**, 106–24.
321. STOCKINGER, K. R., MACKENZIE, A. J. and CARY, E. E. (1963). Yield and quality of sugar beets as affected by cropping systems, *J. Am. Soc. Sug. Beet Technol.* **12**, 492–6.
322. STOJKOVSKA, A. and COOKE, G. W. (1958). Micronutrients in fertilisers. *Chemy Ind.* 1368.
323. STOUT, B. A., SNYDER, F. W. and CARLETON, W. M. (1956). The effect of soil moisture and compaction on sugar beet emergence, *J. Am. Soc. Sug. Beet Technol.* **9**, 277–83.
324. STOUT, M. and SMITH, C. H. (1950). Studies on the respiration of sugar beets as affected by bruising, by mechanical harvesting, severing into top

and bottom halves, chemical treatment, nutrition and variety, *Proc. Am. Soc. Sug. Beet Technol.* **6,** 670–9.

325. SYKES, E. T. (1931). The time for applying nitrate of soda to sugar beet, *J. Minist. Agric. Fish.* 1–11.

326. TERRY, N. (1970). Developmental physiology of sugar beet. II. Effects of temperature and nitrogen supply on the growth, soluble carbohydrate content and nitrogen content of leaves and roots, *J. exp. Bot.* **21,** 477–96.

327. THIELEBEIN, M. (1960). Postulates for germination of sugar beet seed in the field, *Zucker* **13,** 539–45.

328. THORNE, G. N. and WATSON, D. J. (1956). Field experiments on uptake of nitrogen from leaf sprays by sugar beet, *J. agric. Sci., Camb.* **47,** 12–22.

329. TINKER, P. B. H. (1965). The effects of nitrogen, potassium and sodium fertilisers on sugar beet, *J. agric. Sci., Camb.* **65,** 207–12.

330. TINKER, P. B. H. (1967). The effects of magnesium sulphate on sugar-beet yields and its interactions with other fertilisers, *J. agric. Sci., Camb.* **68,** 205–12.

331. TINKER, P. B. H. (1967). The relationship of sodium in the soil to uptake of sodium by sugar beet in the greenhouse and to yield responses in the field, *Proc. int. Soc. Soil Sci.* 223–31.

332. TINKER, P. B. H. (1967). A comparison of the properties of sodium and potassium in the soil, *Inf. Nitr. Corp. Chile* **97.**

333. TINKER, P. B. H. (1970). Fertiliser requirements of sugar beet on fen peat soils, *J. agric. Sci., Camb.* **74,** 73–7.

334. TINKER, P. B. H. (1970). How long does applied sodium remain in the soil? *Inf. Nitr. Corp. Chile* **115.**

335. TOLMAN, B. (1943). Sugar-beet seed production in Southern Utah, with special reference to factors affecting yield and reproductive development, *U.S. Dep. Agric. Tech. Bull.* 845.

336. TOLMAN, B., JOHNSON, R. and GADDIE, R. S. (1956). Comparison of CO_2 and $NaHCO_3$ as extractants for measuring available phosphorus in the soil, *J. Am. Soc. Sug. Beet Technol.* **9,** 51–5.

337. TOMLINSON, T. E. (1970). Urea-agronomic applications, *Proc. Fertil. Soc.* No. 113.

338. TRIST, P. J. O. and BOYD, D. A. (1966). The Saxmundham rotation experiments: Rotation I, Rotation II, 1899–1952, *J. agric. Sci., Camb.* **66,** 327–36, 337–9.

339. TRUOG, E., BERGER, K. C. and ATTOE, O. J. (1953). Response of nine economic plants to fertilisation with sodium, *Soil Sci.* **76,** 41–50.

340. ULRICH, A. (1948). Plant analysis as a guide to the nutrition of sugar beets in California, *Proc. Am. Soc. Sug. Beet Technol.* **5,** 364–77.

341. ULRICH, A. (1950). Critical nitrate levels of sugar beets estimated from analysis of petioles and blades, with special reference to yields and sucrose concentrations, *Soil Sci.* **69,** 291–309.

342. ULRICH, A. and HILLS, F. J. (1952). Petiole sampling of sugar beet leaves in relation to their nitrogen, phosphorus, potassium, and sodium status, *Proc. Am. Soc Sug. Beet Technol.* **7,** 32–45.

343. ULRICH, A. and OHKI, K. (1956). Hydrogen ion effects on the early growth of sugar beet plants in culture solution, *J. Am. Soc. Sug. Beet Technol.* **9,** 265–74.

344. ULRICH, A., RIRIE, D., HILLS, F. J., GEORGE, A. G. and MORSE, M. D. (1959). Plant analysis . . . a guide for sugar beet fertilisation. Analytical methods . . . for use in plant analysis, *Bull. Calif. agric. Exp. Stn.* 766.

345. ULRICH, A. (1961). Plant analysis in sugar beet nutrition, *in* 'Plant Analysis and Fertiliser Problems', pp. 190–211. Washington, Am. Inst. Biol. Sci.

346. ULRICH, A. (1964). The relative constancy of the critical nitrogen concentration of sugar beet plants. Plant Analysis and Fertiliser Problems, IV Colloq., 371–91.
347. ULRICH, A. and HILLS, F. J. (1969). Sugar beet. Nutrient deficiency symptoms, *Univ. Calif. Div. Agric. Sci.*
348. VAN BURG, P. F. J., VAN BRAKEL, G. D. and SCHEPERS, J. H. (1967). The agricultural value of anhydrous ammonia on arable land 1963–66, *Tech. Bull. Neth. N.* **3.**
349. VAN LUIT, B. and SMILDE, K. W. (1969). [Boron fertilisation of sugar beets, based on soil analysis], *Rapp. Inst. Bodemvruchtbhd, Haren* **9,** 1–48.
350. VASKHNIL, P. A. and MANORIK, A. V. (1954). [The enrichment of dung and the use of biologically enriched composts], *Agrobiologija* **3,** 24–33.
351. VILAIN, M. and AVRONSART, J. (1967). [Water economy under sugar beet], *Bull. Ass. fr. Étude Sol* **4,** 5–15.
351A. VÖMEL, A. and ULRICH, A. (1963). Leaf analysis for the determination of manganese deficiency in sugar beet. *Z. PflErnähr. Dung.* **102,** 28–45.
352. VON MÜLLER, K., NIEMANN, A. and WERNER, W. (1962). Influence of nitrogen: potassium ratio on yield and quality of sugar beet, *Zucker* **15,** 142–7.
353. WALLACE, T. (1945). Some aspects of mineral deficiencies in farm crops, *Agric. Prog.* **20,** 20–25.
354. WALLACE, T. (1951). 'The Diagnosis of Mineral Deficiencies in Plants by Visual Symptoms.' London, H.M.S.O.
355. WALSH, T. (1970). Towards efficiency in the use of our soils, *Scient. Proc. R. Dubl. Soc.* **2,** 285–327.
356. WARREN, R. G. (1958). The residual effects of the manurial and cropping treatments in the Agdell Rotation Experiment, *Rep. Rothamsted Exp. Stn. for* 1957, 252–60.
357. WARREN, R. G. and JOHNSTON, A. E. (1960). The exhaustion land site, *Rep. Rothamsted Exp. Stn. for* 1959, 230–39.
358. WARREN, R. G. and JOHNSTON, A. E. (1961). Soil organic matter and organic manures, *Rep. Rothamsted Exp. Stn. for* 1960, 43–8.
359. WARREN, R. G., JOHNSTON, A. E. and PENNY, A. (1962). The value of residues of P K fertilisers in soils. Continuous barley site at Woburn, *Rep. Rothamsted Exp. Stn. for* 1961, 58–9.
360. WARREN, R. G. and JOHNSTON, A. E. (1962). Barnfield, *Rep. Rothamsted Exp. Stn. for* 1961, 227–47.
361. WARREN, R. G. and COOKE, G. W. (1962). Comparisons between methods of measuring soluble phosphorus and potassium in soils used for fertiliser experiments on sugar beet, *J. agric. Sci., Camb.* **59,** 269–74.
362. WATSON, D. J. and RUSSELL, E. J. (1943) *et seq.* The Rothamsted experiments on mangolds, 1872–1940. Parts I, II, III, IV(i), IV(ii), *Emp. J. exp. Agric.* **9,** 49–64; **9,** 65–77; **13,** 62–79; **14,** 49–56; **14,** 57–70.
363. WATSON, D. J. (1947). Comparative physiological studies on the growth of field crops. Parts I and II, *Ann. Bot.* **11,** 41–76, 375–407.
364. WEBB, B. C., HARRISON, C. M. and DEXTER, S. T. (1960). The growth of sugar beets in sand cultures fertilised solely with several green manures, *Q. Bull. Mich. St. Univ. agric. Exp. Stn.* **43,** 367–74.
365. WEBBER, J. (1961). Ploughing down fertiliser for sugar beet. Experiments in Yorkshire 1954–57, *Expl. Husb.* **6,** 8–12.
366. WHITE, T. L. (1959). Petiole analysis as a guide to the manuring of sugar beet, *Pl. Soil* **11,** 78–86.
367. WHITEHEAD, A. G., DUNNING, R. A. and COOKE, D. A. (1971). Docking disorder and root ectoparasitic nematodes of sugar beet, *Rep. Rothamsted Exp. Stn. for* 1970, Part 2, 219–36.

368. WHITEHEAD, D. C. (1963). Some aspects of the influence of organic matter on soil fertility, *Soils Fertil., Harpenden* **26**, 217–23.

369. WHITEHEAD, D. C. (1964). Soil and plant nutrition aspects of the sulphur cycle, *Soils Fertil., Harpenden* **27**, 1–8.

370. WIDDOWSON, F. V. and PENNY A. (1967). Results of an experiment at Woburn testing farmyard manure and N, P and K fertilisers on five arable crops and a long ley. I. Yields, *J. agric. Sci., Camb.* **68**, 95–102.

371. WIDDOWSON, F. V., PENNY, A. and WILLIAMS, R. J. B. (1967). Results of an experiment at Woburn testing farmyard manure and N, P and K fertilisers on five arable crops and a long ley. II. N, P and K removed by the crops, *J. agric. Sci., Camb.* **68**, 293–300.

372. WIDDOWSON, F. V. (1971). Personal communication.

373. WILL, G. H. (1961). Magnesium deficiency in pine seedlings growing in pumice soil nurseries, *N.Z. Jl. agric. Res.* **4**, 151–60.

374. WILLIAMS, R. J. B. (1971). Relationships between the composition of soils and physical measurements made on them, *Rep. Rothamsted Exp. Stn. for 1970*, Part 2, 5–35.

375. WILLIAMS, R. J. B. and COOKE, G. W. (1971). Results of the Rotation I experiment at Saxmundham, 1964–69, *Rep. Rothamsted Exp. Stn. for 1970*, Part 2, 68–97.

376. WILLIAMS, W. A. and RIRIE, D. (1957). Production of sugar beets following winter green manure cropping in California: Part I. Nitrogen nutrition, yield, disease and pest status of sugar beets, *Soil Sci. Soc. Am. Proc.* **21**, 88–92.

377. WILLIAMS, W. A., DONEEN, L. D. and RIRIE, D. (1957). Production of sugar beets following winter green manure cropping in California: Part II. Soil physical conditions and associated crop response, *Soil Sci. Soc. Am. Proc.* **21**, 92–4.

378. WINNER, C. (1966). Düngung überdüngung und qualität der Zuckerrübe, *Potass. Symp.* 1966, 89–106.

379. WOOLLEY, D. G. and BENNETT, W. H. (1962). Effect of soil moisture, nitrogen fertilisation, variety and harvest date on root yields and sucrose content of sugar beets, *J. Am. Soc. Sug. Beet Technol.* **12**, 233–7.

380. WRIGHT, M. J. and DAVIDSON, K. K. (1964). Nitrate accumulation in crops and nitrate poisoning of animals, *Adv. Agron.* **16**, 197–247.

381. YASUDA, T., KUSHIZAKI, M., NISHI, H. *et al.* (1968). [Effects of sodium on sugar beets], *Rep. Hokkaido natn. agric. Exp. Stn.* **92**, 45–53.

382. YATES, F., BOYD, D. A. and MATHISON, I. (1944). The manuring of farm crops: some results of a survey of fertiliser practice in England, *Emp. J. exp. Agric.* **12**, 164–76.

383. YATES, F. and PATTERSON, H. D. (1958). A note on the six-course rotation experiments at Rothamsted and Woburn, *J. agric. Sci., Camb.* **50**, 102–9.

384. YOUNG, H. C. (1943). Fertiliser in relation to black root, *Sugar* **38**, 35–6.

385. ZOCCO, A. (1972). Private communication.

Index